CW01151820

Report on China's Cruise Industry

Hong Wang
Editor

Report on China's Cruise Industry

Springer

Editor
Hong Wang
Shanghai International Cruise
 Business Institute
Shanghai
China

ISBN 978-981-10-8164-4   ISBN 978-981-10-8165-1 (eBook)
https://doi.org/10.1007/978-981-10-8165-1

Jointly published with Social Sciences Academic Press

The print edition is not for sale in China Mainland. Customers from China Mainland please order the print book from: Social Sciences Academic Press.

Library of Congress Control Number: 2018931506

© Social Sciences Academic Press and Springer Nature Singapore Pte Ltd. 2018
This work is subject to copyright. All rights are reserved by the Publishers, whether the whole or part of the material is concerned, specifically the rights of translation, reprinting, reuse of illustrations, recitation, broadcasting, reproduction on microfilms or in any other physical way, and transmission or information storage and retrieval, electronic adaptation, computer software, or by similar or dissimilar methodology now known or hereafter developed.
The use of general descriptive names, registered names, trademarks, service marks, etc. in this publication does not imply, even in the absence of a specific statement, that such names are exempt from the relevant protective laws and regulations and therefore free for general use.
The publishers, the authors and the editors are safe to assume that the advice and information in this book are believed to be true and accurate at the date of publication. Neither the publishers nor the authors or the editors give a warranty, express or implied, with respect to the material contained herein or for any errors or omissions that may have been made. The publishers remains neutral with regard to jurisdictional claims in published maps and institutional affiliations.

Printed on acid-free paper

This Springer imprint is published by the registered company Springer Nature Singapore Pte Ltd. part of Springer Nature
The registered company address is: 152 Beach Road, #21-01/04 Gateway East, Singapore 189721, Singapore

Contents

1 **China's Cruise Industry in 2016–2017: Transformation, Upgrading and Steady Development** 1
Hong Wang, Jianyong Shi, Xinliang Ye, Younong Wang and Junqing Mei

2 **Ten Hot Spots of China's Cruise Industry Development During 2016–2017** 49
Hong Wang, Jianyong Shi, Younong Wang, Xingliang Ye, Ling Qiu, Ke Chen, Junqing Mei, Xie Xie, Mingyuan Wu, Ruihong Sun, Guodong Yan, Xia Li, Tian Hu, Taotao Li, ERbing Cai, Shuang Cai, Fengri Wang and Feng Yu

3 **Active Connection to "The Belt and Road" Initiative to Enable the Great-Leap-Forward Development of Shanghai Cruise Economy** 77
Hong Wang, Xinliang Ye and Shuang Cao

4 **The Development and Changes of Cruise Industry in China Over the Decade** 91
Jianyong Shi, Xinliang Ye and Junqing Mei

5 **Research on Asian Cruise Business Climate Index** 109
Ling Qiu, Juan Luo, Linkai Qi and Hong Jiang

6 **Research on Cruise Industry in China: Current Issues and Prospects** 129
Xiaodong Sun and Rongxin Ni

7 **The Economic Contributions of Evaluation Indicator System to Shanghai's Cruise Industry** 143
Huang Huang

8	**Development Path Research on Shanghai's Cruise Supporting Industry**	177
	Guodong Yan, Shuguang Lei, Yanhui Gao, Ying Ye and Yanan Cai	
9	**Research on the Financial Service System for Development of China's Cruise Industry**	195
	Yingkai Yin, Zheng Yu and Zhihui Jiang	
10	**Legal Coordination Difficulties in the Course of Chinalization of Cruise Industry and Coping Measures**	223
	Fangyuan Lv, Zhaobin Pei and Jie Zheng	
11	**Discussion on the Nature and Governance Mode of Cruise Ship Environmental Pollution from the Perspective of Marine Right**	239
	Ruihong Sun, Hong Wang and Xinde Chen	
12	**Analysis on the Development Strategy and Path of China's Domestic Cruise from the Perspective of Differentiated Competition**	251
	Xinliang Ye, Junqing Mei and Yuanqin He	
References		269

Chapter 1
China's Cruise Industry in 2016–2017: Transformation, Upgrading and Steady Development

Hong Wang, Jianyong Shi, Xinliang Ye, Younong Wang and Junqing Mei

1 Development Environment of China's Cruise Industry in 2017

Being a highly internationalized and high-end service industry, the cruise industry is originated in North America and blooms throughout North America and Europe, currently growing in the Asia-Pacific region. As a highly dynamic and internationalized global industry, the cruise industry is greeting a new global industry pattern along with the constantly changing social and economic situation of the world. North America, the largest global cruise market, shows significant HQ economy and leads the advancement of the global cruise industry. Europe has gained obvious superiority in global monopoly of cruise manufacture attributed to its entire and complete industry chain. The Asia-Pacific region, as an emerging cruise market, grows the fastest and shows the greatest potential of development. Although currently ranking the second in the world and being the most promising one according to the society, China's cruise market is still limited by immaturity in development, which needs to closely follow the development trend of the global cruise industry, borrow experiences, paths and patterns from mature international cruise markets, explore the new mode, momentum and path for the development of China's cruise market so as to provide stronger confidence and support for the world.

H. Wang (✉) · J. Shi · X. Ye · Y. Wang · J. Mei
Shanghai International Cruise Business Institute, Shanghai, China
e-mail: wh@sues.edu.cn

J. Shi
e-mail: shijy@sues.edu.cn

X. Ye
e-mail: yexinliang@sues.edu.cn

© Social Sciences Academic Press and Springer Nature Singapore Pte Ltd. 2018
H. Wang (ed.), *Report on China's Cruise Industry*,
https://doi.org/10.1007/978-981-10-8165-1_1

1.1 International Environment: Continual Growth in Spite of Sluggish Economy of the World

1.1.1 Sluggish Global Economy and Increased Development Uncertainty

During 2016, the global economy still faced grave difficulties, with the global trade growth dropped to the rock bottom since 2009. There was still a lack of endogenous power for economic development while the growth foundation was still weak, leading to an economic growth rate lower than expected. The effective market demand of developed countries was at a low level. The structural reform of developing countries faced difficulties in progress, seeing continually increased risks due to geopolitics. Therefore, the low-speed economic development of the world and the multiple risks were never changed fundamentally, and greater complexity in economic development occurred with the rise of protectionism. According to the *World Economic Situation and Prospects 2017* issued by United Nations Department of Economic and Social Affairs in January 2017, the global economic growth rate was 2.2% in 2016, the lowest among the seven years since 2009. As estimated, although there would be an upturn in the global economy in 2017 (the global economic growth is expected to be 2.7% in 2017), the global economic growth is still low in 2017. Regarding the economic development trend of major regions in the world, the U.S.—the greatest economic power of the world—would see an economic growth rate of 1.7% in 2017 while that of the European region would reach 1.8%. China, the largest developing country of the world, may witness an economic growth rate much higher than the global average and developed countries, possibly reaching 6.5%. India expects a growth rate of 7.7% in 2017, higher than that of China. The International Monetary Fund (IMF), however, is optimistic about the global situation of economic development, anticipating an economic growth rate of 3.1% in 2016 and 3.5% in 2017. For 2018, it is still optimistic with an anticipated growth rate of 3.6%. According to IMF, the primary economic growth momentum comes from the emergency markets worldwide and the fact that anticipated economic development of developing countries tends to be better. From January to March 2017, China will expect a growth rate of 6.5% while that of India is 6.1%, both considered high among global emerging economic powers.

As to the international social, political and economic progress of 2017, there are obvious signs of protectionism, nationalism and isolationism, of which 33% international trade measures are against China. Trade protectionism and the Federal Reserve System raising interest rates will significantly inhibit global trade volume and border crossing. In addition, 2017 is a year seeing the change of leadership of many countries. 1/3 economic powers including Germany, Italy and France will see the selection of the supreme leadership, which, to some degree, increases the uncertainty of global economic development. Besides, populism and the trend of counter-globalization are also the stumbling blocks for the global economy and, to some degree, impede the effective restoration of the global economy. According to statistics of *Economic and Social Survey of Asia and the Pacific 2017* issued by U.N. Economic and Social Commission for Asia and the Pacific in May 2017, economic growth rate of developing economic powers of the Asia-Pacific region would reach 5.0% in 2017 and further to 5.1% in 2018. Although the trend of economic development of developing countries in the Asia-Pacific region is not optimistic and the growth would slow down to some degree, the region would still take a lead in global economic growth through steady development.

Facing the stress of economic downturn worldwide, emerging economic powers have become a new momentum for global economic development, especially China, who has implemented more open diplomatic policies and "the Belt and Road" initiative, is playing a more and more important role in this progress and becomes an indispensable engine to lead the global economy to develop in a sustainable and healthy way.

1.1.2 Continual Growth of Cruise Market and Promoted International Status of the Asia-Pacific Region

The Global Cruise Market Scales up Rapidly, and Regional Concentration of FDI Is Still High

In the past decade, the demand of the global cruise tourism market remained high. According to statistics of Cruise Lines International Association (CLIA), from 2004 to 2016, the tourist volume in the global cruise market increased from 13.14 million per year to 24.70 million per year, with a growth of 88%. CLIA analyzed the prospect of the cruise tourism market in 2017 according to latest ship building planning and arrangement, estimating that the total tourist volume of the global cruise market in 2017 could reach 25.80 million passengers, another increase compared with the estimated 25.30 million passengers of December 2016. The global market, except for North America and Europe, maintained rapid growth in both market demand and market share in the past decade. The market demand increased by 266.1% during the decade from 2005 to 2015. See Fig. 1.

Passenger origins are concentrated in the global market. North America is the home to modern cruise tourism where economy is developed, water conditions are ideal and maritime culture is mature. There have been over 50 years since the start of cruise tourism in this area, during which numerous cruise lines have grown up, and this area has become the most developed and vigorous market of global cruise tourism, also a model for developing the industry worldwide. See Table 1.

In 2016, the U.S. was still the largest origin of cruise tourists in the world, with the total annual cruise tourist volume reaching 11.52 million passengers. China, superseding Germany, became the second largest cruise tourism market of the world with a total annual outbound cruise tourist volume of 2.28 million passengers. Germany, the third in the global market, witnessed a total annual cruise tourist volume of 2.02 million passengers. The U.K, ranking the fourth, witnessed a total

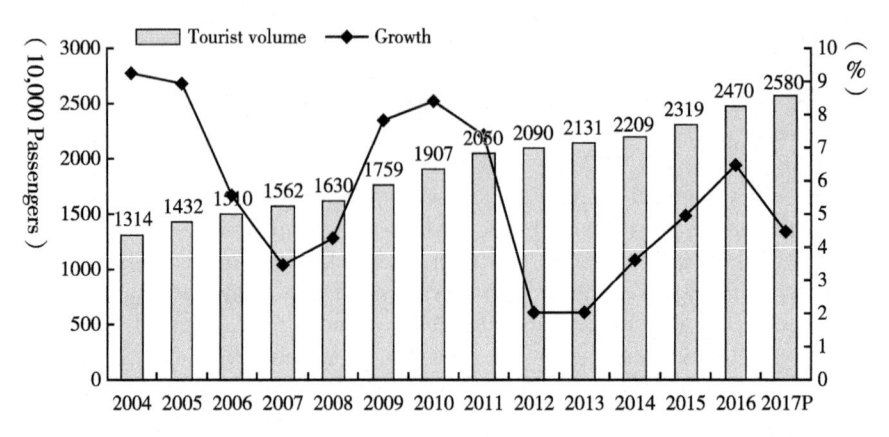

Fig. 1 Development of global cruise tourism market. *Data source* Cruise Lines International Association

Table 1 Demand of the global cruise market during 2005–2015

Region	2005	2009	2010	2011	2012	2013	2014	2015	Decade growth rate
North America	9.96	10.4	11	11.44	11.64	11.82	12.21	12.17	22.1
Europe*	3.15	5.04	5.67	6.15	6.23	6.4	6.39	6.59	109.2
Subtotal	13.11	15.44	16.67	17.59	17.87	18.22	18.55	18.76	43.1
Other regions**	1.21	2.15	2.4	2.91	3.03	3.09	3.49	4.43	266.1
Total	14.32	17.59	19.07	20.5	20.9	21.31	22.09	23.19	61.9

Data source Cruise Lines International Association
*The European market also includes Russia, the Central Europe and the Eastern Europe besides the EU's 27 countries. **Other regions: main estimates and adjustments started from 2009, considering dynamic growth of economy of the Southern Hemisphere
Unit Million Passengers, %

annual cruise tourist volume of 1.89 million passengers. Australia, ranking the fifth, witnessed a total annual cruise tourist volume of 1.29 million passengers. Canada and Italy recorded a total annual cruise tourist volume of 750,000 passengers and the figure of France was 570,000 passengers. Spain and Brazil, of equal amount, witnessed a figure of 490,000 passengers. See Fig. 2.

The most popular areas of cruise tourism in 2016 are the Caribbean (35%), the Mediterranean (18.3%), European regions other than the Mediterranean (11.1%), Asia (9.2%), Australia/New Zealand/the Pacific (6.1%), Alaska (4.2%) and South America (2.5%).

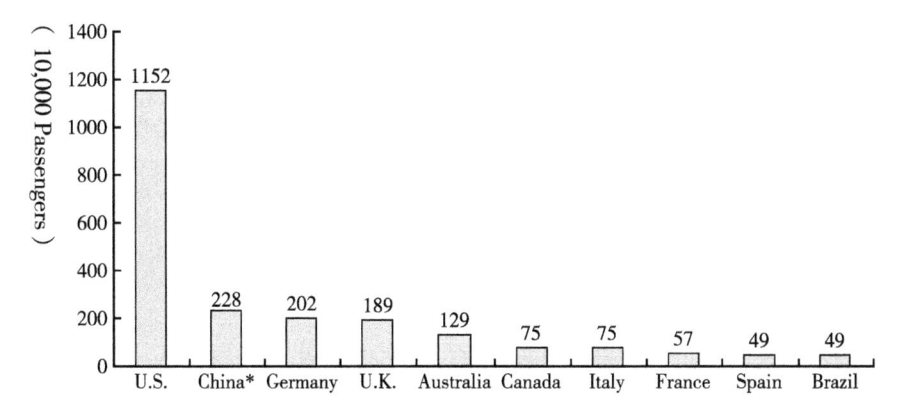

Fig. 2 Top 10 cruise tourism markets of the world in 2016. *Data source* Cruise Lines International Association, *from CCYIA

The International Cruise Tourism Market Sees a Clear Trend of Younger

According to statistics of CLIA, the international cruise tourism market is observed with the following typical features.

(a) The younger generation, including the Millennials and the X Generation (the next generation of the Baby Boom) is more enthusiastic about cruise tourism, whose appraisal for cruise trips are higher than land-based trips, all-inclusive holiday travels, sight-viewing, home tours and camping. Cruise tourists consider cruise tours the best way for mentally relaxing and escaping the pressure.

(b) Tourists who have the experience of cruise tourism show high loyalty to this mode of travel. About 92% passengers expressed that they would probably or definitely take another cruise tour. The younger generation shows the highest loyalty. 2/3 the Millennials and the Y Generation (born during 1980–1995) regard cruise as their favorite mode of travel, in which 56% prefer maritime tours and 10% prefer inland river tours. This feature is even clearer for the X Generation, of which 58% prefer maritime tours and 13% prefer inland river tours.

(c) People feel better when enjoying cruise tours with others, and their appraisals for cruise tours are higher when taking trips with their families or friends.

(d) As to travel planning, people having no experiences in cruise tours are usually more dependent on their companions, who often get advice from friends, family members, destination websites and travel information networks. Tourists who have experiences in cruise tourism, however, obtain information from a wider range of channels including social networks, tour guides, tourism magazines/blogs, APPs and travel agencies.

(e) In comparison with land-based tours, travel agencies can collect more information from cruise tourism marketing and are more willing to provide

assistance to tourists. Tourists are much more satisfied when detailed travel plans are available from travel agencies.
(f) There is a positive correlation between travel time and age of cruise tourist. Although a 6–8 days' period is favorite to people of all ages, young people who follow the fashion trend prefer shorter cruise tours, usually 5 days or less. However, traditionalists prefer cruise tours of long routes, usually 9 days or more.
(g) Despite that most race-based segmented markets show the similarity of consumption preference regarding cruise, black people and African Americans obvious like short-distance cruise tours better, which are usually 5 days or less.
(h) For cruise tours in the future, the most popular destinations will be the Caribbean, Bermuda Islands, Mexico and Northwestern Pacific.
(i) One advantage of cruise tours in comparison with overland travel is that tourists can see and experience new things. Most cruise tourists regard cruise itself as an ideal destination of travel. This is especially true for the Millennials, who prefer more to staying at the visiting port or for a longer time or visiting a destination that is new to them than the older generation.
(j) Besides, there is usually a great gap between facilities that are believed to be significant by tourists and those available for use by them during the tour. For example, tourists often consider the cruise nanny plan important, but seldom adopt it in the tour. On the other hand, many tourists believe that recreational facilities, casino games, shopping spots, gyms, pools and tubs are not so important, which are, however, the most popular facilities used by them.
(k) The greatest factor influencing shopping behaviors of tourists is age. The Millennials value price the most. People born during the Baby Boom consider the port and destination the most important factor. It is interesting that traditionalists regard children and family the most important thing.

The Scale of Cruise Transport Capacity Is Increasing at a High Pace Globally, with Large and New Ships Being Continually Introduced to the World

In the past decade, the deployment of cruise transport capacity keeps a high rate of growth. Bed-days provided by cruise ships around the world increased by 68% in a decade, from 87.30 million in 2005 to 146.7 million in 2015. In 2015, the total bed-days available in North America are 69.60 million, accounting for 47.44% of the total of the world and increased by 19.2% in a decade. In 2015, the total bed-days available in Europe reached 48.80 million, increased by 133.5% in a decade and accounting for 33.26% in total cruise transport capacity of the world. In 2016, there were totally 26 new cruise ships put into operation worldwide, including 9 maritime ships and 17 inland river ships. See Table 2.

In North America, the Caribbean is still the most popular destination of cruise tours. Although the market share dropped from 46.35% in 2005 to 36.2% in 2015, it

Table 2 Development of transport capacity of global cruise market in recent years

Regional market	2005	2010	2011	2012	2013	2014	2015	Decade growth rate
North America	58.4	62.7	62.1	64	63.1	71.7	69.6	19.20
The Caribbean	40.4	46.2	45.5	48	48.1	55.9	53.1	31.40
Other regions in North America	18	16.5	16.6	16	15	15.8	16.5	−8.30
Europe	20.9	41.4	49.5	48.7	49.6	46.4	48.8	133.50
Northern Europe	5.9	9.7	11.4	13.2	13.9	13.5	13.8	133.90
The Mediterranean	15	31.7	38.1	35.5	35.7	32.9	35	133.30
North America + Europe	79.3	104.1	111.6	112.7	112.7	118.1	118.4	49.30
Other regions	8	13.8	15.1	20.7	21.8	24.1	28.3	253.80
Total	87.3	117.9	126.7	133.4	134.5	142.2	146.7	68

Data source Cruise Lines International Association
Unit Million Bed-days, %

remained the status of being the largest destination of cruise tourist worldwide with a transport capacity of 53.10 million bed-days and maintained a growth rate of 31.4% in the past decade and a figure of 5% in 2015. In other regions in North America, mainly Alaska, Mexico, Canada and New Zealand, cruise ship's transport capacity increased by 4.4% in 2015 to 16.50 million bed-days. Despite the gladsome growth, the transport capacity of these regions dropped by 8.3% in 2005.

In Europe, cruise transport capacity has increased by 133.5% during the last decade, from 20.90 million bed-days in 2005 to 48.80 million bed-days in 2015. The growth of cruise transport capacity of Northern Europe and the Mediterranean in the past decade is consistent. However, the Mediterranean witnessed a growth of 6.4% in 2015 while Northern Europe only saw a growth of 2.2% in the same year, which was significantly lower. The transport capacity of Europe deployed in the Caribbean increased from 52% in 2005 to 84% in 2015, accounting for 33% of total transport capacity of the world.

Attributed to powerful drive of Australia and China, the transport capacity of other regions has increased by 254% during the decade since 2005, in which that of 2015 increased by 17%. Transport capacity of regions other than Europe and North America has reached 28.30 million bed-days, holding a proportion of 19% in total cruise transport capacity of the world, which was merely 9% in 2005. In the past decade, the Caribbean and Europe contributed 70–75% of total transport capacity of the world. However, other regions are also experiencing rapid growth.

The large-sized trend of cruise ship is obvious. The market stock for cruise ships greater than 100,000 t is huge. Royal Caribbean International owns numerous large-sized luxury cruise ships, including several over 220,000 t cruise ships, which are the largest of the world. In 2017, the record on the largest cruise ship in the world was broken again by Royal Caribbean International. The "Symphony of the Seas" built by STX France for the company has a tonnage of 230,000, 3000 t greater than the "Harmony of the Seas", the record holder of 2016. The ship has as many as 2775 guestrooms and a total length of 362 m, with 28 more balcony rooms than the "Harmony of the Seas". The ship realizes a standard capacity of 5494 persons and maximum capacity of 6780 persons, with an international crew of 2175 persons.

Royal Caribbean International has placed two 160,000 t cruise ships—"Quantum of the Seas" and "Ovation of the Seas" into the Chinese market. MSC Cruises, a highly reputed international cruise line launched the "MSC Meraviglia" in June 2017, the largest cruise ship ever built by a European ship-owner with a total tonnage of 171,500 t and maximum carrying capacity of 5714 persons. Current, the "MSC Meraviglia" is the fourth largest cruise ship in the world, ranking after three Oasis-class cruise ships of Royal Caribbean International. In 2017, NCLH placed the "Norwegian Joy" into the Chinese market, which as a total tonnage of 167,700 t and carrying capacity of 3850 persons, being the largest cruise ship departing from China. The "Majestic Princess", a cruise ship customized for Chinese tourists, has a total tonnage of 143,000 t and maximum carrying capacity of 3560 persons.

The Economic Contribution of the International Cruise Industry Is Increasing

According to the latest economic contribution statistics of CLIA, in 2015, there were totally 122.7 million passengers (including both tourists and employees) took shore excursions, leading to a total direct expenditure of USD 54.11 billion (USD 58.59 billion with exchange rate adjustment) at the travel destination and in other markets of resources, which included direct expenditures on procurement of products and services of cruise lines to support their operations. These expenditures produced an output (direct, indirect and induced) of USD 117.9 billion (USD 126.97 billion with exchange rate adjustment). Totally 956,500 employment opportunities were created and a total revenue of USD 38.16 billion was achieved (USD 40.76 billion with exchange rate adjustment). According to CLIA's statistics, the revenue of the global cruise market in 2016 reached USD 42.79 billion, increased by 10.6% a year ago. See Fig. 3.

The rapid growth of the global cruise market is mainly attributed to the development of the Asian market, which is an emerging and fast-growing section of the global cruise market. In 2016, the cruise carrying capacity in Asia increased by 38% from 2015, holding a proportion of 9.2% of the global cruise market.

The Asian Cruise Market Grows Vigorously with Increased Global Influences

In 2016, cruise carrying capacity of Asia held a proportion of 9.2% in the global cruise market, i.e. 2.2724 million passengers, increased by 38.1% from 2015 and

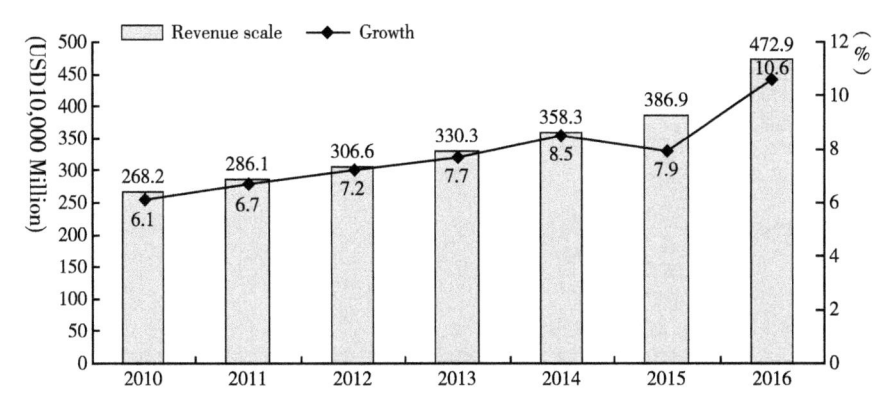

1 China's Cruise Industry in 2016–2017: Transformation … 9

Fig. 3 Revenue scale of global cruise market. *Data source* Cruise Lines International Association

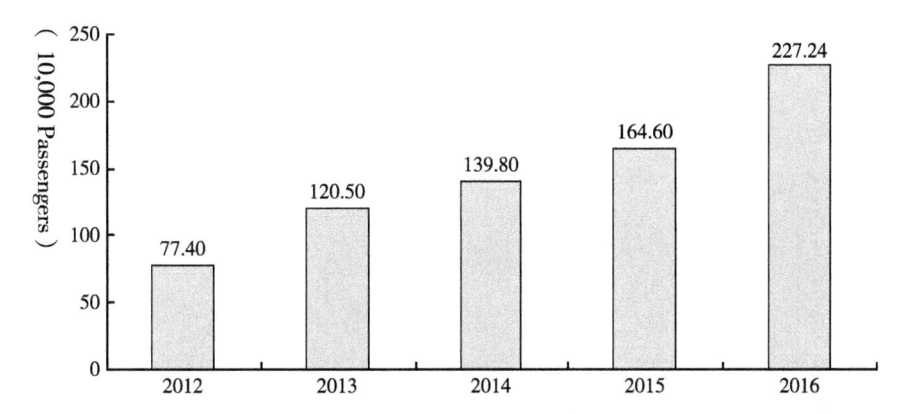

Fig. 4 Changes in passenger volume direction of Asian cruise market. *Data source* Cruise Lines International Association

193.6% from 2012. In recent years, the Asian market experienced rapid growth: 55.7% from 2012 to 2013, 16.6% from 2013 to 2014 and 17.7% from 2014 to 2015, being the fast-growing cruise market of the world. See Fig. 4.

Regarding the passenger volume direction, 47.4% passengers are from China, the most of Asia, and 11.1% from Taiwan, 8.8 from Singapore, 8.6% from Japan, 6.1% from Hong Kong, 6.0% from India, 3.0% from Malaysia, 1.9% from Indonesia, 1.7% from South Korea, 1.6% from Philippines, 1.2% from Thailand and 0.9% from Vietnam. 40% of passengers are aged below 40. It is clear that China is the largest origin of cruise passengers in China, while Japan and South Korea, the largest tourist destinations of Asia, only contribute a small proportion of cruise passengers, less 10% in total and even smaller than the proportion of Taiwan (11.1%). See Fig. 5.

Regarding cruise transport capacity, there were totally 31 cruise brands in operation in the Asian market in 2016. The number was 26 in 2015. The total

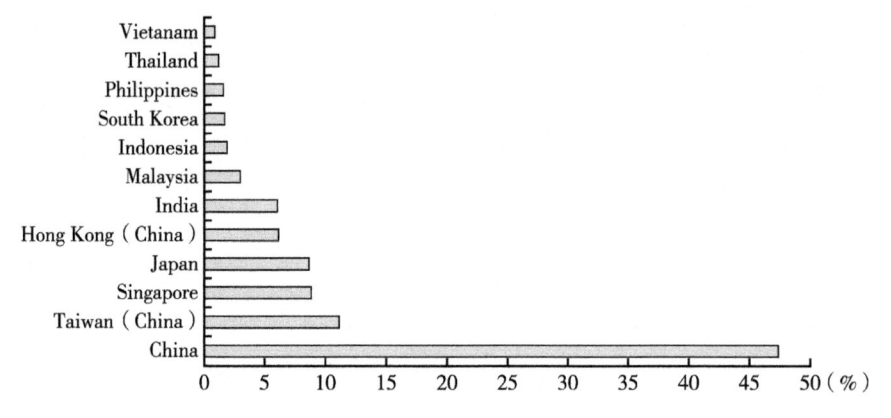

Fig. 5 Changes in origin of cruise passengers of Asian cruise market. *Data source* Cruise Lines International Association

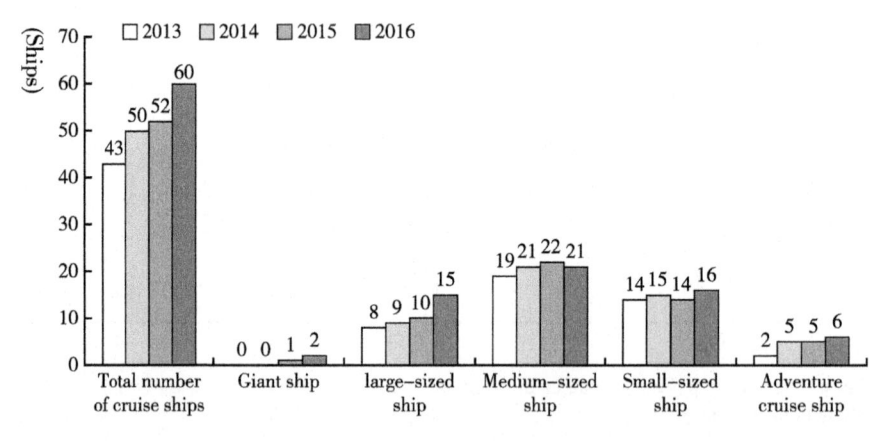

Fig. 6 Changes in type of cruise ships of Asian cruise market. *Data source* Cruise Lines International Association

number of cruise ships in operation reached 60, with a total number of voyages of 1560, total number of inbound berthing of 5500 ships, total number of days of port operation of 7918, including 204 destinations. The number of potential passenger destination-days reached 10.90 million and passenger volume reached 3.1 million, which was 1.4, 1.51 and 0.775 million separately in 2014, 2013 and 2012. See Fig. 6.

Regarding the type of ships placed in the market by international cruise lines, among the 60 ships of 2016, there were 21 medium-sized ships, 15 large-sized ships, 16 small-sized ships, 6 adventure cruise ships and 2 giant ships. The percentage of giant ships is still very low, mainly due to the crowdedness, operational cost and difficulties in operation of extra-large ships. The percentage of medium and large-sized ships reached 60%, showing an obvious large-sized trend of cruise ships that is mainly due to the popularization of cruise tour. Large carrying capacity of ship can satisfy diversified leisure demands of passengers and provide more catering and recreational facilities and create a comfortable feeling in a broad space. Besides, larger carrying capacity means more passengers and more revenue, also highlighting the attribute of cruise ship to be the destination of a trip. However, over-crowded giant ships are not favorable for good cruise experiences.

Regarding cruise products offered by cruise lines, most products are 4–6-night cruise tours (48%), followed by 2–3-night tours (32%), 7-night tours (6%) and open-sea tours (5%). In general, the longer the trip is, the less the quantity of cruise product will be. According to the development trend of cruise products during 2013–2014, 2–3-night short-period cruise products had a large proportion (over 40%) but showed a trend of gradual decline, reduced by 8% point in 2016 to 32%. The percentage of 4–6-night cruise products was rising gradually to 48% in 2016 (by 14% point), mainly because the 2–3-night schedule is too tight to achieve the objective of travel. 4–6-night products provide enough time for passengers to experience cruise tourism at a relatively rational price, this percentage has increased gradually. The selection for 7-night tour also increased to some degree, but the percentage is still low. Other longer schedules are in small proportions, directly related to the holiday system and travel expenditure. See Fig. 7.

Inbound berthing: refer to the record of inbound berthing instead of the calling of a cruised ship, including the departure from and return to the home terminal. Regarding the volume of inbound berthing of cruise ships, in 2016, Japan saw 1526 ships, ranking the first in the Asian market, followed by China (850 ships, Shanghai the mostly frequently visited terminal), South Korea (745 ships, Jeju the largest terminal), Vietnam (466 ships), Malaysia (422 ships, Penang the largest terminal), Singapore (391 ships), Thailand (291 ships, Phuket Island the largest terminal), Taiwan (234 ships, Taipei and Keelung the largest terminals), Hong Kong (185 ships) and

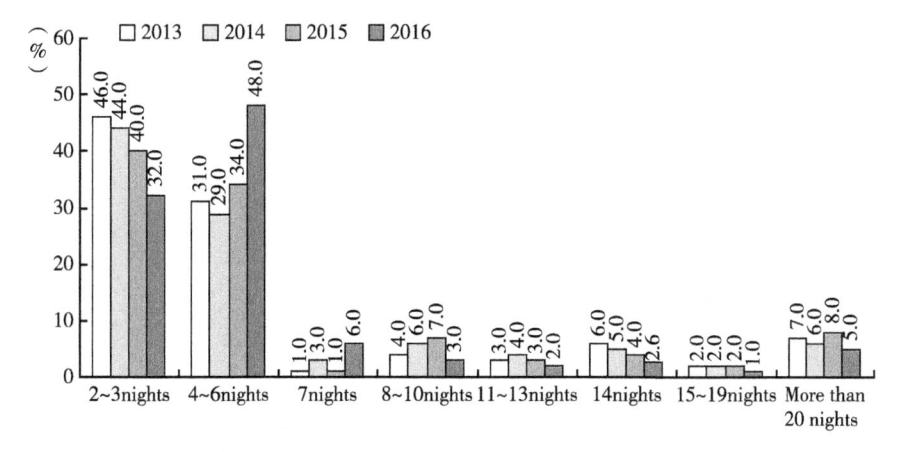

Fig. 7 Changes in cruise products of Asian cruise market. *Data source* Cruise Lines International Association

Indonesia (172 ships) in sequence. The figure shows that Japan, South Korea and China receive the largest amount of inbound berthing in Asia. See Fig. 8.

The Global Cruise Building Market Is Highly Concentrated and Shows Obvious Trend of Monopoly

Currently, cruise ship manufacturing shipyards are mainly distributed in Europe including Italy, Germany, Finland, France, etc. As of the middle of June 2017, there were totally 74 luxury cruise ships in ordering state which were distributed in 19 shipyards. Among which, the 15 European shipyards held the orders of 69 cruise ships, meaning a 98% share of the global market by gross tonnage. Regarding specific numbers, there are 6 shipyards owned by Fincantieri that have 27 in-process ship orders. Meyer Werft GmbH of Germany owns 2 shipyards and has 19 in-process orders. STX France has 7 in-process orders. In addition, Alstom Marine of France and Aker Finnyards of Finland also have different quantities of orders. For the moment, matching products of STX France have realized a localization rate of 94% in Europe. The rest 6% products are from Asia or other regions. Products of Fincantieri have a European localization rate of 99%, while Meyer has realized a European localization rate of 98%.

Fincantieri of Italy is a state-owned shipbuilder and one of the largest shipbuilders of Europe, who has powerful technical advantages and rich experiences in the design and building of luxury cruise ships. Fincantieri, being the largest luxury cruise ship builder worldwide, has a share of more than 30% in the global luxury cruise market (by historical building performance). The group totally owns 6 merchant ship

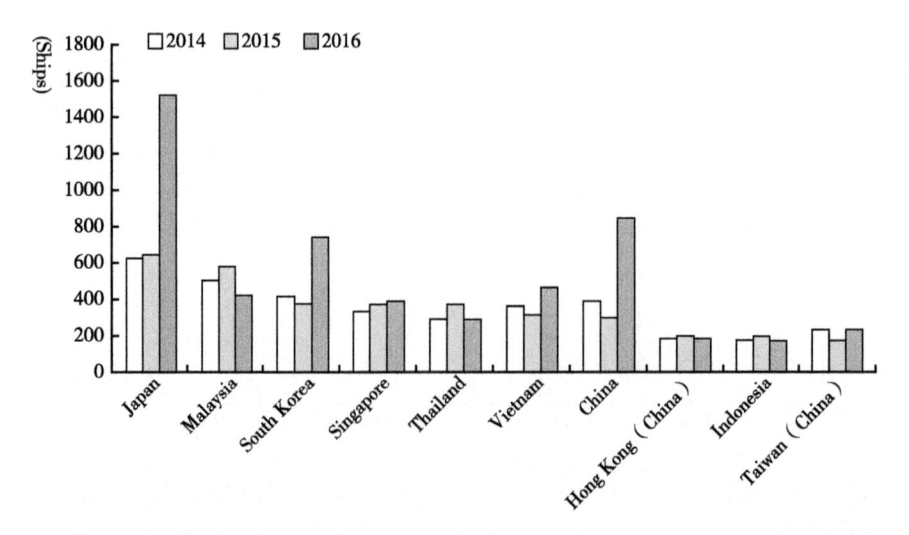

Fig. 8 Trend of cruise ship inbound berthing volume of Asian cruise market. *Data source* Cruise Lines International Association

construction yards, of which Monfalcone and Marghera shipyards are the main builders of large-sized luxury cruise ships.

At present, main businesses of Fincantieri include building extra-large luxury cruise ships and small-sized inner river ships. The rated seat of an extra-large luxury cruise ship is generally greater than 3000 with main customers of Carnival Cruise Lines and MSC Cruises. The rated seat of a small-sized inner river ship is usually less than 1000 with main customers of Viking Cruises and some American cruise lines. As of the middle of June 2017, Fincantieri's shipyards had held orders of 27 cruise ships with a total tonnage of 2.433 million ton, holding a proportion of 33.4% in total in-process orders of the world's luxury cruise ships (by tonnage). France has a highly developed shipbuilding industry. STX France, the former Chantiers de l'Atlantique, was first built in 1861, when 33% shares were held by the French government. The company totally built 120 cruise ships during the past 150 years. On April 6, 2016, STX French and MSC Cruises signed a letter of intent for 2 + 2 luxury cruise ship building which had a total value of approximately EUR 4 billion (about USD 4.5 billion). MSC Cruises has totally subscribed 12 cruise ships from STX French (including those already delivered) since 2003. STX French built the world's largest luxury cruise ship—"Harmony of the Seas"—for Royal Caribbean Cruises Ltd (227,500 t, carrying 6000 passengers and a crew of 2100).

STX French is the only money-spinning enterprise subordinate to STX Group that is engaged in building large-sized luxury cruise ships. In the past, the company's performances were mainly focused on 3000–4000-seat extra-large cruise ships, with customers of Royal Caribbean Cruises Ltd, Star Cruises and other world-famous cruise lines. Till today, the company has in-process orders of totally 7 cruise ships with a total tonnage of 1.154 million tons, holding a proportion of 15.8% in total in-process orders of the world (by tonnage).

As of the middle of June 2017, STX France, Fincantieri and Meyer separately held 15.8% (1.154 million ton), 33.4% (2.433 million ton) and 40.0% (2.911 million ton) of in-process orders worldwide. Since then, Fincantieri will occupy half of the market and show an absolute and dominant advantage.

STX France owns more abundant production resources. As to the time of delivery of in-construction luxury cruise ships, all 7 ships of STX France can be completed by January 2020. Among the 27 ships constructed by 6 holding enterprises subordinate to Fincantieri, 11 can be delivered after 2020. Among the 19 ships constructed by Meyer, there are 8 scheduled to be delivered after 2020. Once new orders come, STX France will have enough production resources to meet new market demands.

Germany is one of the countries who have mastered advanced technologies on cruise ship building, and luxury cruise ships and passenger ships are the only one place for Germany's shipbuilding industry to take a lead globally. According to statistics of the end of 2015, over 61% of all shipbuilding products of Germany are luxury cruise ships and passenger ships. Germany is also the main supplier of ship supporting products of China, South Korea and Japan, three largest origins of shipbuilding in the world. German shipyards developed against the global trend and witnessed an order amount of EUR 5 billion in 2015, almost doubled compared with 2013. Order amount in 2016 reached the same level as 2015. The success of Germany's shipbuilding industry is attributed to technical advantages and high reliability of ship products, which are difficult to follow by other countries in a time span of dozens of years. The "Norwegian Joy" ordered by Norwegian Cruise Line and built by Meyer was customized for the Chinese market.

Meyer Werft GmbH of Germany, a well-known specialized luxury cruise ship builder is the largest shipbuilder of the country, which was founded by the Meyer family in 1795. At the beginning, Meyer only built small wooden ships, and then it started the business of iron ships from 1874. Up to now, Meyer has successively delivered over 700 ships of various types including ferryboats, roll-on/roll-off boats, liquefied gas carriers, container carriers, livestock transport ships, inner river cruise ships, ocean liners, oceanographic vessels, etc. With the rise of Asian shipbuilding industry which has taken a great majority of orders for ordinary civilian ships, Meyer began to transfer its business focus to the building of luxury cruise ships.

Meyer has been mainly engaged in building passenger ships and ferryboats for a long time. It started in the market of cruise ship building in middle 1980s attributed to its experiences and technologies related to passenger ship and luxury ferryboat building, and succeeded in building the first cruise ship—the "Homeric". Since then, Meyer's business began to focus on building 3000–4000-seat extra-large cruise ships. For the moment, Meyer has the orders of 19 cruise ships with a total tonnage of 2.911 million ton, holding a proportion of 40.0% in total in-process orders for luxury cruise ships of the world (by tonnage). Although the number of ships in building is less than Fincantieri, the total tonnage is obviously greater than the latter.

Lloyd Werft was founded in 1857, once an old shipyard frequently engaged in cruise ship modification and maintenance and only built new and small-sized inner river cruise ships and some special ships. Since the 1970s, Lloyed Werft transformed to develop cruise ship modification and new ship building. The investment and restructuring of Genting Hong Kong integrated the shipyard with neighboring shipyards to enter into the market of large-sized luxury cruise ship building.

1.2 Domestic Environment: New National Strategies Bring Along New Opportunities for Rapid Development of the Tourism Industry

1.2.1 "The Belt and Road" Initiative Brings Along New Opportunities for Development of Cruise Economy

"The Belt and Road" initiative will advance China's inbound and outbound tourism into a new prime period. "The Belt and Road" initiative, especially the framework of the 21st Century Maritime Silk Road, offers a wider space for routing design of China's cruise industry. According to authoritative estimate of the China National

Tourism Administration (CNTA), China will totally realize 150 million Chinese tourists and over USD 200 billion tourism expenses for countries along "the Belt and Road". "The Belt and Road", in all directions centered on China, gets through the channels to Europe, the Mediterranean and the Indian Ocean. In particular, the 21st Century Maritime Silk Road starts from Chinese ports, pass through the South China Sea and the Indian Ocean and arrive at Europe, getting through the channels from China to Europe, the Mediterranean and other regions where the cruise market develops the best, being an extension of China's international, intercontinental and interoceanic routes, making countries along "the Belt and Road" the largest destinations of China's cruise market and providing an enormous space for development of China's cruise industry.

During the past decade, terminal-based cruise ships became the backbone for the rapid development of the cruise market. With the pace of development of the terminal-based cruise ship slowing down, inbound cruise tourism will become a new market momentum. According to authoritative estimate of CNTA, during the 13th Five Year Plan Period, China will attract a total of 85 million passengers from countries along "the Belt and Road" and realize a total tourism consumption of USD 110 billion approximately. To advance inbound tourism, China has implemented preferential policies including "144-h visa-free transit" and "15-day visa-free entry". The wider range of regions applying free-visa policies, the extension of the visa-free period, the greater effort made in international promotion of China's cruise tourism and destinations and the setup of overseas promotion agencies have enhanced China's national image and attraction to countries on "the Belt and Road". China's inbound cruise tourism will further developed with the promotion of the Maritime Silk Road high-quality cruise tourism route and the establishment of the local brand (the Silk Road) international cruise tourism.

The development of the cruise industry requires a strong professional support. At present, there is a lack of professional cruise professionals in China and it is difficult to provide senior cruise talents in the short term. The suggestion is to introduce more professionals from overseas cruise industries. With the implementation of the strategy of "the Belt and Road", the foreigners' understanding of China becomes more profound. As the maturity of the cruise market in Europe and the United States has been gradually matured and gradually decreased growth rate, the world's new cruise market has gradually increased, and the attraction to the cruise talents in the western countries will gradually increase. More and more talents will participate in the development of the Asia-Pacific cruise market.

The macroscopic economy develops steadily and progressively, and the achievements of the supply-side structural reform start to show. Under the guidance of the new development concept, China's economic development has shown a steady trend, and the development potential of economic growth mainly comes from the good market development, the new drive brought by innovation and the industrial structural adjustment due to economic transformation. Current, the macroscopic economy is developing within a rational range, featured in higher sustainability, stability, growth rate and coordination. In 2017, the supply-side structural reform will go on in a comprehensive manner to realize qualitative and quantitative growth and the ultimate healthy development of China's economy. In 2016, China's GDP reached 74.4127 trillion yuan, with a year-on-year growth of 6.7%, of which the service industry showed ideal development trend with the proportion continuing to increase, reaching 3.84221 billion yuan, with a year-on-year growth of 7.8%, accounting for more than 50% (reaching 51.6%), realizing a year-on-year growth of 1.4%. In 2016, China's total retail sales of social consumer goods reached 33.2316 trillion yuan, seeing a year-on-year growth of 10.4% from a year ago. To remove the impact of price changes, the year-on-year growth was 9.6%. Foreign economic growth maintained a good momentum, with the number of imports and exports in 2016 reaching 5.3484 trillion yuan, with a year-on-year growth of 14.2%, and import and export deficit reaching 1.7097 trillion yuan. In the first half of 2017, China's GDP reached 38.1490 trillion yuan, seeing an increase of 6.9% over the previous year according to comparable prices. The added value generated by the service sector reached 20.6516 trillion yuan, with a year-on- year growth of 7.7%. See Fig. 9.

China's residents' income level and spending power keep rising. In 2016, China's per capita disposable income reached 23,821 yuan, seeing a year-on-year growth of 8.4%. To remove the impact of price changes, the year-on-year growth was 6.3%. The median per capita disposable income is rising, reaching 20,883 yuan, with a year-on-year growth of 8.3%. The per capita disposable income of urban residents reached 33,616 yuan, with a year-on-year growth of 7.8, and 5.6% to remove the impact of price changes. The median per capita disposable income of urban residents reached 31,554 yuan, with a year-on-year growth of 8.3%. The countrywide per capita consumption expenditure was 17,111 yuan, with a year-on-year growth of 8.9% from a year ago. The increase without price influence was 6.8%. The per capita consumption expenditure of urban residents reached 23,079 yuan, with a year-on-year growth of 7.9%. To remove the impact of price changes, the year-on-year growth was 5.7%. In the first half of 2017, China's per capita disposable income of residents reached 12,932 yuan, with a year-on- year growth of

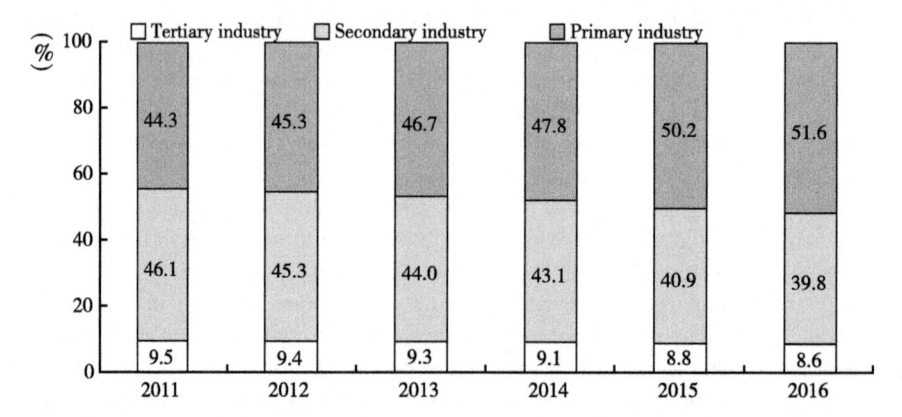

Fig. 9 Changes in proportions of three sectors in GDP during 2011–2016. *Data source* Statistical Bulletin of the State Statistics Bureau

8.8%. To remove the impact of price changes, the year-on-year growth was 7.3%. The per capita disposable income of urban residents reached 18,322 yuan. To remove the impact of price changes, the year-on-year growth was 6.5%. Improved residents income level and spending power lay a solid foundation for leisure-oriented tours and great inner demand potential for the tourism industry. According to the Economic Blue Book's analysis of China's economic development situation in 2017, in the first quarter and the second quarter of 2017, the GDP growth rate will be 6.5 and 6.5%, while in the third quarter and the fourth quarter, the GDP growth rate will reach 6.4 and 6.4%. In 2017, the annual GDP growth rate will reach 6.5%, with a decrease of 0.2% point on a year-on-year basis.

1.2.2 The Tourism Industry Develops Fast, Increasing the Demand for Both Domestic and International Tourism

In recent years, China has realized rapid socio-economic development and people's living standards continued to improve. Mere material life can no longer meet the pursuit of high quality of life. Tourism, as an important part of spiritual enjoyment, has gradually come into people's lives. At present, the rapid development of China's tourism industry has enabled the tourism industry to fully integrate into the national strategic system and become a strategic pillar industry of the national economy. The scale of outbound tourist volume and the tourism consumption are the highest in the world, and the cooperation with various regions and international tourism organizations has been continuously strengthened. In 2016, China's domestic tourism tourists reached 4.4 billion passengers, a year-on-year growth of 11.2%, with the domestic tourism revenue reaching 3.939 trillion yuan, a year-on-year growth of 15.25%. The inbound tourism market continued to grow rapidly, reaching 138.44 million passengers in 2016, increased by 3.5% on a year-on-year basis, and the number of inbound foreign tourists was 28.13 million, increased by 8.3% on a year-on-year basis. In addition, the number of tourists from Hong Kong, Macao and Taiwan reached 110.31 million, with a year-on-year growth of 2.3%. International tourism keeps a high rate of growth. In 2016, the scale of outbound tourism of Chinese residents reached 135.13 million passengers, with a year-on-year growth of 5.75%. Tourists from Hong Kong, Macao and Taiwan were still the absolute majority of the outbound tourism market which reached 83.95 million passengers, decreased by 2.2% on a year-on-year basis. The revenue from international tourism is continuously increasing, reaching USD 120 billion in 2016, with a year-on-year growth of 5.6%. See Fig. 10.

The state vigorously promotes the development of inbound tourism and has taken positive measures in the duty-free shop examination and approval for construction, foreign tourists entry visa, border inspection and air traffic rights and other aspects, promote from the "single-point breakthrough" to "coordinated progress" change for the provision of a strong security system for the rapid and steady development of inbound tourism. In 2017, China has received a total of 69.5 million foreign tourists on inbound tours. The growth rate of the outbound tourism

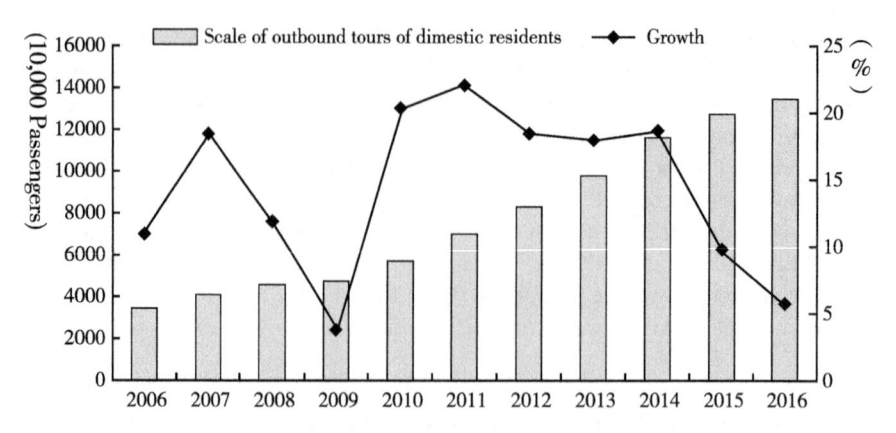

Fig. 10 Changes in the outbound tourism market of China during 2006–2016. *Data source* Statistical Bulletin of the State Statistics Bureau

market has transformed from rapid to steady growth. In the first half of 2017 (January to June), the size of the outbound tourism market of Chinese residents has reached 62.03 million passengers, and the size of the inbound tourism market continues to be greater than that of the outbound tourism market, with a difference reaching 7.47 million passengers. More and more foreigners reacquainted themselves with China along with the rapid development of inbound tourism. China, a country of oriental beauty, is becoming an important tourist destination in the global tourism market.

The construction of a well-off society will greatly increase people's demand for leisure-oriented tourism. With the promotion of global tourism, "tourism +", supply-side structural reform and paid vacation system, China's tourism industry will experience fast growth. In such a new phase and period, high-quality tourism consumer demand will continue to be met, which will further scale up China's tourism market.

2 Development Status of China's Cruise Industry in 2017

2.1 The Cruise Market Continues to Scale up, Leading to a Highly Improved International Status

With over ten years' development, China's cruise market has grown greatly and become the largest section of the Asia-Pacific region. The Asia-Pacific market consists of three main regions: Southern Pacific regions represented by Australia, New Zealand and Indonesia, Southeastern Region represented by Malaysia, Philippines, Singapore and India and Far East region including China, Japan, South Korea and North Korea, in which China is the most active region in the Asia-Pacific cruise market and is becoming the new center of the global cruise market.

According to statistical data of China Cruise & Yacht Industry Association (CCYIA), eleven Chinese cruise ports including Shanghai, Tianjin, Dalian, Qingdao, Zhoushan, Wenzhou, Xiamen, Guangzhou, Shenzhen, Haikou and Sanya received a total of 1181 cruise ships in 2017, with a year-on-year growth of 17%. In terms of the types of cruise ships received, the number of home port cruise ships amounted to 1098, with a YoY growth of 18%; the number of visiting port cruise ships amounted to 83, almost the same as that in 2016. With regard to the number of cruise tourists received, these eleven Chinese cruise ports received 4,954,000 inbound and outbound tourists in total, with a YoY growth of 18%. Specifically, the number of home port Chinese inbound and outbound tourists totaled up to 4,289,700, increased by 93% compared with 2015, and this is the first time that such number exceeds 2 million since 2006. The number of overseas cruise tourists entering China totaled up to 277,500, with a YoY growth of 8%. Among the Chinese top ten cruise ports, the Shanghai Cruise Port received a total of 512 cruise ships and 2,978,000 inbound and outbound cruise tourists, accounting for 43.3% of the domestic market share and 60.1% of the total national number respectively. The Tianjin International Cruise Home Port received up to 175 cruise ships and 942,000 inbound and outbound cruise tourists, accounting for 14.8% of the domestic market share and 19% of the total national number respectively. The cruise market has developed by leaps and bounds in Guangzhou: in 2017, the number of cruise ships received was up to 122 and the number of inbound and outbound cruise tourists reached 403,000, accounting for 10 and 8% of the total national number respectively. See Table 3 for details.

Table 3 Comprehensive analytical statistics of Chinese major cruise ports 2017

Cruise port	Number of cruise ships received in 2016	Number of cruise ships received in 2017	YoY growth (%)	Number of tourists received in 2016	Number of tourists received in 2017	YoY growth (%)
Shanghai	509	512		289.4	297.8	
Tianjin	142	175	23	71.4	94.2	31
Sanya	25	12	−52	9.6	4	−58
Xiamen	79	77	−3	19.0	16.18	−22
Qingdao	52	63	21	8.9	10.94	27
Zhoushan	13	15	15	1.7	3.06	73
Dalian	27	31	15	6.4	6.9	6
Guangzhou	104	122	17	32.5	40.1	23
Haikou	41	33	−20	6.4	2.56	−60
Shenzhen	14	109	679	4.4	18.9	325
Yantai	4	0	–	0.5	0	–
Total	1010	1181	17	456.7	495.4	18

Source China Cruise & Yacht Industry Association

2.2 The Cruise Ship fleet Is Expanding and the Cruise Market Witnesses fierce Competition

In 2016, main brands operating in China's cruise market were the "Ovation of the Seas", the "Quantum of the Seas", the "Voyager of the Seas", the "Mariner of the Seas" and the "Legend of the Seas" owned by Royal Caribbean Cruise Ltd., the "Costa Serena", the "Costa Atlantica", the "Costa Victoria" and the "Costa Fortuna" owned by Costa Cruises, the "Sapphire Princess" and the "Golden Princess" owned by Princess Cruises, the "MSC Lirica" owned by MSC Cruises, the "SkySea Golden Era" owned by SkySea Holding International Ltd., the "MSC Splendida" owned by Diamond Cruise International, the "Chinese Taishan" owned by Bohai Ferry Co., Ltd., the "Libra" and the "SuperStar Virgo" owned by Star Cruises and the "Genting Dream" owned by Dream Cruises, totally 18 home-port cruise ships, which was 12 in 2015.

Regarding regional distribution, ships homing in Shanghai included the "Quantum of the Seas", the "Costa Serena", the "MSC Spendida", the "Sapphire Princess", the "Costa Fortuna", the "MSC Lirica", the "Skysea Golden Era", the "Mariner of the Seas" and the "Chinese Taishan"; ships homing in Tianjin included the "Ovation of the Seas", the "Costa Atlantica" and the "Golden Princess"; ships homing in Qingdao included the "Chinese Taishan" and the "Legend of the Seas"; ships homing in Xiamen included the "Libra", the "Sapphire Princess" and the "Voyager of the Seas"; ships homing in Guangzhou included the "SuperStar Virgo" and the "Genting Dream"; ship homing in Shenzhen included the "Silver Shadow"; ships homing inn Dalian included the "Legend of the Seas" and the "MSC Lirica".

In 2017, there are 18 home-port cruise ships operating in the Chinese market, including 4 ships owned by Royal Caribbean Cruises Ltd. (1 less from a year ago): the "Ovation of the Seas", the "Quantum of the Seas", the "Voyager of the Seas" and the "Mariner of the Seas"; 4 ships owned by Costa Cruises: the "Costa Serena", the "Costa Atlantica", the "Costa Victoria" and the "Costa Fortuna"; 1 ship owned by MSC Cruises: the "MSC Lirica"; 1 ship owned by SkySea Holding International Ltd.: the "Skysea Golden Era"; 1 ship owned by Diamond Cruise International: the "MSC Spendida"; 1 ship owned by Bohai Ferry Co., Ltd.: the "Chinese Taishan"; 2 ships owned by Princess Cruises: the "Sapphire Princess" and the "Majestic Princess"; 1 ship owned by Star Cruises: the "SuperStar Virgo"; 1 ship owned by Dream Cruises: the "Genting Dream"; 1 ship owned by Norwegian Cruise Line: the "Norwegian Joy" and 1 ship owned by Silversea Cruises: the "Silver Shadow".

Princess Cruises will completely withdraw from the Chinese market in 2018, who have been a cruise brand highly praised by the Chinese cruise society and dedicated to the medium and high-end markets. The "Majestic Princess", which was customized to the Chinese market, will withdraw from the Chinese market in 2018 after operation some time in China.

Looking at the geographical location of home-port cruise ships at China's major terminals in 2017, Shanghai will still be the region showing the strongest transport capacity of the cruise market. However, Tianjin in the north and Guangzhou and

Shenzhen in the south are developing stronger market vitality and attraction and becoming the new areas of placing cruise ships by major international cruise operators. Currently, the Eastern China market shows a higher penetration rate than other emerging regions. Developed economy, water conditions, advantages in passenger source and preferential policies jointly contribute to the change in the structure of China's home-port cruise market. At present, 90% cruise passengers are new customers. There is still great pressure relying on repurchase of existing customers, which is an opportunity for emerging markets.

There are totally 12 home-port cruise ships placed in East China, of which 10 home at Shanghai Wusongkou International Cruise Terminal, including the "Costa Serena", the "Costa Atlantica" and the "Costa Fortuna" owned by Costa Cruises; the "Sapphire Princess" and the "Majestic Princess" owned by Princess Cruises; the "Skysea Golden Era" owned by SkySea Holding International Ltd.; the "Quantum of the Seas" and the "Mariner of the Seas" owned by Royal Caribbean Cruise Ltd.; the "Norwegian Joy" owned by Norwegian Cruise Line and the "SuperStar Virgo" owned by Star Cruises. One home-port cruise ship at Shanghai Port International Cruise Terminal: the "MSC Spendida" owned by Diamond Cruise International. One home-port cruise ship at Zhoushan Islands International Cruise Terminal: the "Chinese Taishan" of Bohai Ferry Co., Ltd.

There are 5 home-port cruise ships placed in the North China market, including 3 berthing at Tianjin International Cruise Terminal: the "Costa Fortuna" owned by Costa Cruises, the "MSC Lirica" owned by MSC Cruises and the "Ovation of the Seas" owned by Royal Caribbean Cruises Ltd. One home-port cruise ship at Dalian International Cruise Terminal (in planning): the "Costa Victoria" owned by Costa Cruises. 2 home-port cruise ships at Qingdao Cruise Terminal: the "MSC Spendida" owned by Diamond Cruise International and the "Costa Victoria" owned by Costa Cruises.

In South China market, there are 9 home-port cruise ships, 3 at Guangzhou Port International Cruise Terminal: the "Genting Dream" and the "World Dream" owned by Dream Cruises and the "Costa Victoria" owned by Costa Cruises. 2 home-port cruise ships at Shenzhen Merchants Shekou International Cruise Terminal: the "SuperStar Virgo" owned by Star Cruises and the "Silver Shadow" owned by Silversea Cruises. 3 home-port cruise ships at Xiamen Cruise Terminal: the "Chinese Taishan" of Bohai Ferry Co., Ltd., the "Costa Victoria" owned by "Costa Cruises" and the "Skysea Golden Era" owned by SkySea Holding International Ltd. One home-port cruise ship at Haikou Nanhai Mingzhu International Cruise Terminal (in construction, currently using Port of Xiuying): the "Chinese Taishan" of Bohai Ferry Co., Ltd.

2.3 Preferential Policies Are Introduced to Support the Manufacture of Home-Made Cruise Ships and the Era of Manufacturing Cruise Ships Starts in China

Cruise ships are the core of development of the cruise industry and the most important means of staying power in the market. Governments at all levels have provided vigorous support for manufacturing home-made cruise ships to complete the cruise industry chain of the country and improve economic efficiency of the industry. On October 12, 2016, CSSC International Cruise Industry Park was officially founded in Shanghai of China, aiming at providing manufacturing, operational and other supportive services for building luxury cruise ships and completing the cruise industry chain. See Table 4.

Table 4 Policies related to China's building of home-made cruise ships

Administrative authority	Policy
The State Council	Made in China 2025
	Outline of the National Strategy of Innovation-Driven Development
	Outline of the 13th Five-Year (2016–2020) Plan of the People's Republic of China
	Circular of the State Council on Printing and Issuing the Overall Plan for the China (Zhejiang) Pilot Free Trade Zone
	Circular of the State Council on Issuing the Plan for Deepening the Reform and Opening-up of the China (Shanghai) Pilot Free Trade Zone in All Aspects
	Circular of the State Council on Issuing the Plan for Tourism during the 13th Five-Year Plan Period
	Several Opinions of the State Council on Promoting the Reform and Development of the Tourism Industry
	Several Opinions of the State Council of the People's Republic of China, on Promoting the Sound Development of the Shipping Industry
	Several Opinions of the State Council of the People's Republic of China, on Further Promoting Tourism Investment and Consumption
	Guidance of the State Council on Promoting the Cooperation in International Productivity and Equipment Manufacture
	The 13th Five-Year Plan for Maritime Economy Development of China
Ministry of Transport	The 13th Five-Year Plan for Integrated Transport Services
	Action Plan for Deepening the Supply-side Structural Reform in the Field of Water Transport (2017–2020)
Ministry of Industry and Information Technology	Action Plan to Deepen and Accelerate the Transformation and Upgrading of the Ship Building Industry 2016–2020

(continued)

Table 4 (continued)

Administrative authority	Policy
	Opinions on Promoting the development of the Tourism Equipment Manufacture Industry
Shanghai Municipality	The 13th Five-Year Plan for Promoting the Development of the High-end Equipment Manufacture Industry of Shanghai Municipality
	Executive Summary of Made in China 2025 by Shanghai Municipality
	Planning for Building Shanghai International Shipping Center during the 13th Five-Year Plan Period
	Opinions of the Shanghai Municipal Government on Promoting the Supply-side Structural Reform
	The 13th Five-Year Plan for Transformation and Upgrading of Shanghai's Manufacturing Sector
	Implementation Plan for Pilot Projects in Service and Trade Innovation Development of Shanghai Municipality
	The 13th Five-Year Plan for Developing the Service Sector of Shanghai Municipality
Tianjin Municipality	Outline of the 13th Five-Year Plan for Economic and Social Development of Tianjin Municipality
	Implementation Plan for Building the North International Shipping Core Area of Tianjin Municipality
	Implementation Opinions on Accelerating the Implementation of the State's National Free Trade Zone Strategies
Qingdao City	Circular on Issuing the Implementation Plan for the National Tourism Reform and Innovation Pilot Area of Qingdao City
	Outline of the 13th Five-Year Plan for Economic and Social Development of Qingdao City
	Outline of the 13th Five-Year Plan for Economic and Social Development of Shenzhen City
Xiamen City	The 13th Five-Year Plan for Integrated Transportation Development of Xiamen City
Sanya City	The 13th Five-Year Plan for Maritime Economy Development of Sanya City
Guangzhou City	Several Provisions on Accelerating the Development of Guangzhou's International Cruise Industry

From the central government to local governments, ideal policy supports are provided for building home-made cruise ships. According to Made in China 2025, it is clearly mentioned that the building of cruise ships is the key and it is necessary to enhance the development of core technologies for cruise ship building. According to Several Opinions of the State Council on Promoting the Reform and Development of the Tourism Industry, it is clearly stated that the government will provide vigorous support for domestication of tourism equipment manufacture and the development of cruise tourism. According to Several Opinions of the State Council of the People's Republic of China, on Further Promoting Tourism Investment and Consumption, it is stated that great support will be provided to ensure rapid development of the cruise industry as well as in obtaining core technologies for building large-sized luxury cruise ships, ship assembly and supportive facilities for ship building, and ensure conditions are provided for shipbuilding enterprises owning advanced technologies to research and develop large and medium-sized cruise ships. In addition, plenty of financial support will be granted for enterprises engaged in tourism equipment innovation so as to accelerate technical innovation of the industry.

According to Action Plan to Deepen and Accelerate the Transformation and Upgrading of the Ship Building Industry 2016–2020 issued by *the Ministry of Industry and Information Technology*, support will be provided for building luxury cruise ships: initiate the building of luxury cruise ships as major special projects; make breakthroughs in basic and building technologies of cruise ship manufacture; accelerate the progress of independent design and construction of luxury cruise ships. In addition, support will also be provided for building design of yachts. According to Opinions on Promoting the development of the Tourism Equipment Manufacture Industry, it is proposed to improve building design of home-made cruise ships and accelerate the cooperation between Chinese enterprises with strong power in ship building with overseas enterprises taking a lead in the R&D and design of cruise ships. Efforts will be made to complete the supportive facilities for building luxury cruise ships so as to perfect the standard system for independent construction of home-made cruise ships. The goal is that, within 5–10 years, China could basically obtain the R&D, manufacture, repair and other core technologies related to medium and large-sized cruise ships, build an integral chain for luxury cruise ships consisting of domestic general assembly enterprises, cruise supportive facility service enterprises and cruise ship decoration enterprises and play an important role in international cruise ship design and building.

2.4 China's Home-Made Cruise Industry Is Beset with Difficulties and Huge Operation Stress

As numerous international cruise lines are busy contending for the Chinese market, keywords including "localized cruise ship", "local cruise line", "local cruise tourism", "made in China cruise ship", "off-shore cruise tour" have been popular topics in the cruise society. China's terminal-based cruise economy started in 2006, when the market was dominated by foreign cruise ships and there were few market access limits for foreign cruise lines. The Chinese government supported foreign cruise lines and encouraged the development of the cruise market. As a result, Costa Cruises, Royal Caribbean Cruise Ltd. and other foreign cruise lines placed their cruise ship fleets in the Chinese market, forming the current situation of the Chinese cruise market which is dominated by foreign cruise lines.

With the market growing rapidly, some influential Chinese enterprises marched into the cruise market, including HNA, CTRIP, Bohai Ferry Co. Ltd., who currently only operate one ship, and the overall disadvantage of local cruise ships of China is still huge, producing low profit, even suffering severe loss. In November 2015, the "Henna" retired, while the "Skysea Golden Era", the "Chinese Taishan" and the "MSC Spendida" are still in operation. SkySea Holding International Ltd. is a "local cruise line" having the best operating performances in China, who still faces huge operating stress in the competition with international giants. The "Skysea Golden Era" can be called as the "national cruise ship", the most popular one among Chinese tourists and honored as "the best local cruise ship". However, its operation and management are in hands of Royal Caribbean Cruises Ltd. and the inner decoration is still of the international style, for which it can hardly be called a real "local cruise ship", but more like a "localized cruise ship". At present, local cruise enterprises also operate the Japan-South Korea routes in direct competition with foreign cruise giants. The lack of differentiation leads "cost performance" to be the best selling point of local cruise lines.

In a market constantly eyed by foreign competitors, local cruise lines are disadvantaged in ship tonnage, the use of second-hand ship, low brand value, weak financial power, poor operation and management ability and the lack of cruise talents, which are hard to be solved in a short period. Although China has absolute strengths in distribution channels of cruise ship tickets, the "chartered" mode of operation has caused great impact on distribution channels of the Chinese cruise market. The price competition is brutal, resulting in continually decreased ticket price. The price war and the "chartered" mode of operation has caused some impact to foreign cruise operators as well, but the influence on local cruise lines and travel agency-based distributors is obviously more fierce, leading to significant reduction in revenues of the latter, while foreign operators can make up for their losses from overseas markets, leading to a gradually enlarged gap in competitiveness between local and foreign cruise lines.

2.5 Overseas Shopping Is Still the Focus and There Is a Lack of In-Depth Experience in Cruise Tours

In today's cruise market of China, onshore shopping is "an important part" or "a critical part" of many tourists to participate in their cruise tours. Shopping is an important part of tourist spending, even more important in cruise tours. Many experienced cruise tourists were deeply impressed by the spectacular scene at a duty-free shop. The moment getting off the tourist bus, people swarmed to the duty-free shop. Many even believed it the greatest regret of the cruise tourism not to visit the duty-free shop, and the mood in the tour was damaged. Especially on trips to South Korea and Japan, many middle-aged and elderly people returned to the ship both hands full.

According to tourism big data, in 2016, the spending during outbound tours of Chinese tourists were USD 109.8 billion (760 billion yuan), meaning a per capita expenditure of USD 900, and the figure of the South America route was even higher: over 50,000 yuan. The significant increase in outbound tourist spending aggravated the unfavorable exchange and the escape of domestic demand. Then, is spending of outbound tourists well planned or purely on impulse? During tours to Japan and South Korea, many tourists believed that quality of foreign products were higher, e.g. cosmetics of South Korea and longevity medicine, health care products and home appliances of Japan. Some of them had even made a list of shopping before the travel, making the cruise tourism a totally shopping trip. The main cause of the high outbound tourist spending is the absolute trust of consumers in "made overseas". Research shows that consumers can always find the psychological hint that "made in Japan" commodities are superior to domestic products in price, quality, functionality and brand effect, even showing the absolute trust in "made in Japan" commodities. Research also shows that the middle-young age groups with more cultural capital tend to shop as planned or nearly planned regarding outbound tourist spending, who can obtain more information about products on the net.

This phenomenon reflects the immaturity of China's cruise market in the current phase and the relatively low cognition for cruise tours by tourists. Many tourists never considered the ship as the destination, but always discussed about shopping during the trip. Some did online homework for the commodities to buy instead of the way of properly spending time on the ship. Some argued that onshore shopping constitutes an important part of revenues of travel agencies. In a market with low cruise ship ticket prices, travel agencies could get commissions by guiding tourists to specific duty-free shops. This, undoubtedly, is one reason for high outbound spending of Chinese tourists. As a matter of fact, no matter which travel mode is adopted, large amounts of shopping expenditures are inevitable for tours to Japan and South Korea. In addition, shopping is also an important aspect of cruise economy. However, we shall never put the cart before the horse, and spiritual experience shall outweigh the pure shopping behavior during a trip. In the cruise market, travel agencies often promote various tour packages to tourists. For example, LY.com has promoted a great variety of packages to tourists, including special shopping tours and non-shopping tours, in which shopping tours are usually given as a free gift that is popular among the tourists for onshore touring.

According to overseas expenditure statistics of Bank of China, airport duty- free shops, large shopping malls and luxury shops are where Chinese tourists spend the most overseas. According to overseas expenditure statistics of Bank of China, Chinese tourists spend the most in the United States, with a per capital bankcard consumption more than 18,000 yuan. Outbound tourist spending is a periodical characteristic. However, the consumption in tours to Japan and South Korea nowadays is merely a small portion of overall outbound tourist spending. Most cruise tourists purchase daily supplies rather than luxuries or high-value products with a specific plan made in advance. Some tourists, however, especially women, will still buy on impulse. It is always a grandeur that cruise tourists swarm to a relatively small duty-free shop for shopping at the same time. Think about the crowdedness and the bustle. However, it is still a periodical characteristic, which will be replaced by more harmonious scenes with more and more participation of the younger generation.

In the coming years, shopping will still be an important recreation that is hard to be given up by many tourists. Nevertheless, with the market becoming mature, tourists will turn their attention to the leisure attribute of cruise tours, and more tourists in pursuit of travel quality will choose better onshore travel destinations.

2.6 The Consciousness of Cruise Ship as the Travel Destination Is Weak and There Is a Long Way to Go to Realize Cultural Building

We can clearly see the significance of cruise route in China's cruise market behind the THAAD event. Currently, Chinese tourists still consider the "destination" more important than "the cruise ship", and lack the consciousness that "the cruise ship is the destination of trip". As believed by many Chinese tourists, the value of "travel" is to get to an onshore "destination", not the trip onboard, reflecting a years' awareness of rapidly reaching the "destination" via fast means of traffic. Most Chinese tourists have not formed such a consciousness that the cruise ship itself is the "destination of travel", most believing that the cruise ship is nothing but a more luxury and comfortable means of traffic. In the current Chinese market, tourists can enjoy lots of services and products at a relatively low cost. Such high cost performance is surely welcomed by tourists. We are always emphasizing that the cruise ship is an "offshore resort" and an "offshore five-star hotel". In fact, the cruise ship is still a relatively restricted space and crowded in comparison with an "overland resort". A 70,000 t cruise ship like the "Skysea Golden Era" can at most accommodate 1814 passengers. The "Quantum of the Seas"—the largest cruise ship of Asia—has a total tonnage of 167,000 t and maximum carrying capacity of 4180 passengers. Activity areas are restricted in these relatively constrained spaces.

Therefore, many cruise tourists, when talking about their cruise tours, will usually say "I took a trip to Japan/South Korea by cruise ship" instead of "I took a cruise tour". Tourists are much more focused on the destination than the cruise ship itself. The question asked most frequently is "where are the cruise ships traveling to?" instead of "what are the most favorable cruise ships to be recommended?" For a particular cruise ship, questions are usually focused on the cost performance of the ship. There is still a long way to go to educate the current market. The innovative change of people's life style is revolutionary, and the cultivation of such life patterns is very fast, e.g. Didi Taxi, Mobike, etc., which can be widely promoted throughout the country within a short period of time. However, cruise tourism is not a "rigid demand" in life, although travel will gradually become one. Cruise tourism itself is still a specific tourism product featured in great price elasticity of demand and is still far from being a "rigid demand product".

2.7 The International Political Environment Is Complex and Tricky and the Weakness of Cruise Tourism Is Highlighted

Tourism is a service industry that is highly correlated to other factors and influenced by international and domestic politics, economic development and social culture. Therefore, the dramatic change in one of the influencing factors may trigger the domino effect and lead to significant variations in the development trend. At present, outbound tourism is the most popular form of cruise tourism in China, which is inevitably exposed to "geopolitics". The international relationship with neighboring countries will also influence China's cruise industry to a certain extent. Public policies play a critical role in the development of cruise tourism. The guidance of policies can produce significant effects. Manpower, goods and financial resources will automatically pivot towards areas supported by public policies.

A cruise tour is a complete process: the origin—offshore cruising—the onshore destination—the origin, of which the "onshore destination" is inevitably subject to the entry policies of the country or region of destination. Considering the centrality and batching characteristics of cruise tours, rapid customs clearance has become an important node of reform. Many countries have introduced policies towards rapid clearance and visa-free entry to promote local tourism. Complexities in entry formalities, complicated clearance procedures and even the rejection of entry of tourists from certain countries/regions will directly affect the destination preferred by a tourist and the change in the travel plan. Another critical concern is the impact of geopolitics on China's cruise market. An ideal geopolitical relationship is key to fast and healthy development of China's cruise market. However, the relationship between China and countries of destination of cruise tourism are completed and variable, bringing along many uncertainties for the cruise market of China.

3 Puzzles Faced by China's Cruise Industry in 2017

3.1 The National Development Plan Needs to Be Innovated and Improved

In recent years, the international cruise tourism in China has maintained a good momentum driven by strong demand for taking relaxing vacations and supported and guided by governments at all levels, of which the growth is must faster than the

European and North American areas where the international cruise tourism has experienced a long history of development and accumulated lots of experiences. Currently, China's cruise tourism market is experiencing fast growth. However, the driving effect on regional economy of the cruise industry fails the anticipation, and the economic contribution is relatively low in comparison with the international cruise industry. At the moment, the main development pattern of China's international cruise tourism is that Chinese tourists take luxury cruise ships owned by foreign cruise lines to visit Japan, South Korea, etc., and the number of foreign tourists visiting China by cruise ship is still small, that is, there are fewer inbound cruise ships calling at ports in China. Foreign cruise lines have been playing a key and dominating role in China's cruise market, taking the most profitable segment in the cruise industry chain. Cruise operation revenues mainly include those from ticket selling, tourists' onboard spending and spending at the destination, most of which flow to headquarters of foreign cruise lines. At present, China is engaged in only a few profiting segments in the international cruise tourism and the profitability is weak. Most operation revenues are from berthing fees of cruise terminals, fees related to cruise ship provisioning, price spread of travel agencies, etc., at relatively low level in comparison with revenues of international cruise lines. Although China has developed several international cruise terminals in several regions, there is still a lack of global cruise tourism development plan as well as the guidance of the government in developing the cruise tourism. In an overall context of extremely unbalanced development among regional cruise markets, in the first three quarters of 2017, many cruise terminals received less than 30 cruise ships and some even did not see a single cruise ship calling in a whole month. Therefore, the healthy development of China's cruise industry needs a national cruise tourism plan that provides instructions for the development orientation and key fields of local cruise markets, so that regional markets are interlinked and unnecessary competitions are avoided. The international cruise tourism is a highly global industry, of which the effective allocation of the transport capacity of cruise ships is the key to promoting the development of the industry. The cruise tourism market needs to develop in a more balanced way by setting up the market access system for international cruise lines and introducing incentive policies to regional markets.

3.2 China's Cruise Route Policies Call for Innovation

International cruise tourism, as a concept and consciousness of an emerging travel destination, has not matured in a rapidly growing market like China's. Most tourists are more interested in the cruise terminal and consider it a key to their trips. For the moment, restricted by location and revenue management and other factors, there are fewer options for cruise trips departing from China. The lack of options will, to some degree, damage the market attraction and reduce the willingness of existing cruise tourists to take more trips in the future, which is unfavorable for the sustainable and rapid growth of the international cruise market. Among outbound cruise routes departing from China, those to Japan and South Korea occupy an important position. 95% outbound cruise ships departing from China are to Japan and South Korea. The problem is that the routes to Japan and South Korea are subject to the political ties of China with these countries. Any change of international relations will bring along severe consequences for China's cruise market. Other destinations currently available are farther from China's cruise terminals. Such trips will take longer time and economic cost of tourists. Thus, China's cruise market needs to expand the range of operation and explore more cruise routes.

There are complexities in approval routes to Hong Kong, Macao and Taiwan and some restrictions are imposed on visas and passes to these regions. Due to the uncertainties in domestic cruise routes, numerous terminals, coastal resources and featured tourism resources are hard to integrate with cruise tourism, and the cruise tourism routes in the offshore area are not developed. Routes to the South China Sea are currently in the exploration phase. The lack of use of the rich and unique maritime resources of the South China Sea results in a relatively small-scale cruise market in this region. Cruise tourism in this region is severely restricted by the lack of abundant route options.

The THAAD event itself only influences the routes to South Korea, and in fact promotes the development of the routes to Japan. China's cruise market is increasing popular among tourists, for whom the "lack of abundant routes" has become a major concern. For the moment, the Japan/South Korea routes have occupied almost more than 95% of main operational routes of the market, from international cruise lines such as Royal Caribbean Cruises Ltd., Costa Cruises and Prince Cruises to domestic cruise lines such as SkySea Holding International Ltd. and Bohai Ferry Co., Ltd. Such market situation causes concerns about the growth and scale of the tourist market: will tourists' willing to visit Japan/South Korea be constant? Can Japan/South Korea routes guarantee rapid growth of China's cruise market? How long can such attraction last for? Particularly speaking of the destinations in Japan and South Korea, main destinations in Japan are Fukuoka, Sasebo, Nagasaki, Kagoshima, Kumamoto, Kochi, Kitakyushu, Okinawa and Hiroshima, and those in South Korea are Jeju Island, Inchon, Busan, etc. Obviously, there are more destinations in Japan.

Departing terminals of Japan/South Korea routes mainly include Dalian, Tianjin, Qingdao, Yantai, Shanghai, Zhoushan and Xiamen, while Guangzhou and Shenzhen mainly operate routes to Vietnam, Hong Kong, Japan and South Korea. The cruise market in China obviously heavily relies on Japan and South Korea routes, considering geographical location, operational cost and travel time/cost of tourists. Cruise lines hope to see lower operational cost while tourists want to view more places at a relatively low cost. Chinese tourists are not willing to spend too much time onboard, thus Japan and South Korea, which are relatively close to mainland China, become the most favorite options.

The rely on Japan and South Korea routes not only limits the diversified demand of Chinese tourists, but, to some degree, constraints the scale of China's cruise market and damages its capability to resist risks. Once the demands for these routes shrink or there are changes of international relationships, the operation cost (in particular areas north of Xiamen) will increase significantly. At present, there are few regions around China that can become ideal destinations or berthing terminals for cruise tours. Japanese and South Korean terminals can be called at within 48 h from many cruise ports. However, such locational advantage is obviously relative. It will take much longer time from Tianjin, Dalian and Shanghai to the Southeast Asia. Besides, it is very unlikely that more routes to Taiwan will be developed.

3.3 The "Multi-port Call" Business Policy Needs to Be Innovated

To further promote the development of the cruise tourism market and interlinked developed of coastal cities as cruise terminals, the Ministry of Transport had issued the Circular on the Special Permission for Multi-Port Call of Foreign Cruise Ships in China in October 2009, stated that foreign cruise ships were allowed to make multi-port call in China with proper approval. However, in the past few years, it was found that such policy barely had any effect on promoting the domestic cruise market and failed the anticipation. This is significantly related to the strict limit on approval. According to the Circular, the multi-port call of foreign cruise ships are categorized as domestic transport, of which the operation must be specially permitted by relevant authorities. It is a long and complex procedure to obtain such "special permission", including the submission of basic information about the cruise ships applying for "multi-port call", the berthing agreement and other required application materials, with each application handled specially according to the situation. However, the Circular clearly stated that passengers should not stay at the port of call and must return to the ship after disembarking.

The purpose of the "multi-port call" policy is to better utilize the abundant cruise terminals of China, alleviate the problems existing in ports of coastal regions (the lack of operation and poor interaction of passenger source), enhance all-round cooperation in the cruise tourist market, provide higher convenience for tourists to take part in cruise tours and achieve balanced growth of the cruise market. However, in reality, very few foreign cruise lines had promoted multi-port call routes since the introduction of the policy in 2009. This is not only because of the market itself, but also due to the complexities in and the long period for the approval procedure and the request that "passengers should not leave the ship without returning". All these factors have made the "multi-port call" policy less practical in reality.

3.4 The Cruise Ship Provisioning Policy Calls for Innovation

Provisioning of cruise ships guarantees the success of cruise tours and necessary material supply for daily operation of cruise ships, also an important consideration of improving economic efficiency of an international cruise terminal. In the Asian cruise market, international cruise lines often make the provisioning of cruise ships in Japan and South Korea. China is lagged behind in the field of cruise ship provisioning mainly due to the policy constraints in this regard. International cruise lines usually carry out global purchase or global purchase + local transport depending on product quality and cost control requirements. For the moment, China has not officially introduced any policies towards the provisioning of cruise ships, and the policies on general cargo ships are followed. The provisioning of cruise ships are not included into the general trading customs declaration system, so that the provisioning of ships which is a rather simple practice abroad becomes extremely complex in China. Currently, the provisioning of cruise ships has not integrated with national standard practices in terms of export rebates, customs clearance for traded goods and international container transfer. The supervision procedures are numerous and excessively strict, with higher taxes imposed on the provision of cruise supplies than the international market. China has not established any special comprehensive bonded zones or warehouses for cruise ships, or any particular procedure of approval and supervision for this industry. Besides, China has imposed strict conditions for cruise ship provisioning service providers and not permitted local cruise lines to take part in the process of cruise ship provisioning yet, so that local cruise lines can hardly gain more market profit in this field and become restricted in expanding local cruise business. In terms of services related to cruise ship provisioning, some regions have intended to extend the policies on FTZs to international cruise terminals. However, such process kept jogging on and the result failed the anticipation.

3.5 New Policies Towards the Purchase of Cruise Ships Are Needed

In recent years, China's cruise market has experienced rapid growth and local operators have been trying to enter the cruise industry. However, there haven't been any Chinese cruise ships in operation so far. The "Skysea Golden Era" owned by SkySea Holding International Ltd. subordinated to CTRIP and the "HNA- Henna" owned by HNA (retired upon expiration) are registered in Malta; the "Chinese Taishan" of Bohai Ferry Co., Ltd. is registered in Panama, and the "MSC Spendida" owned by Diamond Cruise International is registered in the Bahamas. Till now, China still has no cruise ships that are registered in China and hang the five-star national flag. All cruise ships at present are registered in Panama or the

Bahamas as foreign cruise ships and hang the flag of convenience, which is to a large extent related to a 30% tax. Local cruise ships should have hung the national flag. However, in reality, all these cruise ships are registered overseas and hang the flag of convenience mainly due to the high tax on ships (VAT, import duties) and policy constraints. Tax is one of the main causes that most local cruise ships prefer to be registered overseas. According to Chinese tax regulations, if a local company purchases a foreign cruise ship and register it in China, the import VAT is imposed at a rate of 17%, together with the import duties of 9%. In actual operations, cruise ships registered in China need to pay some other taxes and fees including corporate income tax, vehicle and vessel use tax, stamp tax and other taxes. Another reason for overseas registration is related to the operation management and control of cruise ships. A cruise ship hanging the Chinese national flag shall have a crew of whom 80% are Chinese, a contradiction with the nature of internationalization of the cruise industry that restrains the diversity of cruise ship operations.

Chinese cruise ships are prohibited from lottery-related business, even on international waters, which will, to some degree, restrain business operation of the ship. There are strict procedures for the approval of cruise ship duty-free shops. There are also problems relating to ship approval. In China, there is a certain quota for international ships hanging the Chinese national flag. The approval procedure is quite strict and the process is tedious. Besides, ships hanging the Chinese national flag shall unconditionally follow international shipping policies of the Chinese government, presenting great potential political risks. Last but not the least, China implements extremely strict ship inspection system which, to some extent, forces local cruise operations to adopt overseas registration.

According to current policies, there are no particular provisions on the service life of cruise ships and those on ferries are adopted. Provisions on the service life of ferries indicate that no cruise ships aged over 30 years are allowed to operate in the Chinese market, which shall be subject to mandatory scrapping. Nevertheless, cruise ships are high-end vessels that integrate a great variety of high-end technologies, the "pearl of the shipbuilding industry", and the collective creation of high-class technologies of the shipbuilding industry. Cruise ships are greatly different from traditional ro-ro passenger ships, and, according to international practices, the service of a cruise ship shall be much longer than an ordinary ro-ro passenger ship. A cruise ship having operated for ten years is exactly in her prime age. Even those having operated 20–30 years are in the prime of their lives. Luxury cruise ships aged 40–50 years are very popular in Europe and North America where the industry develops into a mature stage. Thus, the limit of 30 years for the service life of a cruise ship will lead to accelerated depreciation of cruise ships owned by local cruise lines during the operation and thus a great amount of depreciation cost is expected, while foreign cruise lines can place their 30 years' old cruise ships into overseas market to avoid the risk of mandatory scrapping, thus creating huge competitive stress for local players.

3.6 Considerable Challenges Are Faced by the Domestic Cruise Industry

With the implementation of "the Belt and Road" national development strategy, the maritime power strategy and the supply-side structural reform and guided by the 13th Five-Year Plan for the tourism industry of the country, the maritime tourism economy welcomes major development opportunities. Cruise tourism, as an integral part of the state's maritime economy, one of the key drive-forces to achieve the maritime power strategy and one pivot for realizing transformation of port cities, will gradually grow into a key momentum that advances the transformation and upgrading of the state's shipping industry, maintains rapid growth of maritime economy, promotes tourism-based diplomacy as scheduled and continues to improve the status of China's tourism in the world. Cruise tourism has gradually become an important part of China's tourism and an important engine that drives the globalization, internationalization and brand marketing of the tourism industry. Local cruise tourism is the destination of China's cruise tourism progress that creates a wider imaginary space for developing the cruise market in China and gradually grows into the main force and leader of the market.

China has a long coastline extending from the Yalujiang Estuary in the north to the Beilun Estuary in the south, a total of more than 18,000 km. Coastlines of islands reach a total of over 14,000 km, including more than 400 km natural deep-water lines. China's waters include the Bohai Sea, the Yellow Sea, the East China Sea, the South China Sea and waters east of Taiwan, of which the Bohai Sea has a mainland coastline of over 2700 km, and those of the Yellow Sea, the East China Sea and the South China Sea are more than 4000, 5700 and 5800 km respectively. Abundant coastlines are the exceptionally unique natural conditions for developing local cruise tourism.

Coastal cities of China including Dalian, Qingdao, Shanghai, Xiamen, Shenzhen, Sanya, etc. are well-known tourist destinations that not only own abundant tourism and port resources, but also the main origins of passengers of the cruise market, creating ideal conditions for developing offshore cruise tours. The South China Sea area has wide waters and pleasant natural scenes, where hydrological conditions are ideal for developing local cruise tourism. The strong attraction of the South China Sea to local cruise lines lies in the strong interest from tourists, protective policies, great advantage in maritime resources and the vast and wonderful underwater world.

In spite of these abundant resources, the local cruise industry is facing multiple difficult situations and bottlenecks. The "Skysea Golden Era", the "Chinese Taishan" and the "MSC Spendida" operating in the Chinese market are Chinese-funded, but not local, cruise ships. There is an urgent need to build local cruise ships so as to develop local cruise tourism. There are complexities for these cruise ships to hang the Chinese national flag, which are often registered overseas. The "Skysea Golden Era" is registered in Malta; the "Chinese Taishan" is registered in Panama, and the "MSC Spendida" is registered in the Bahamas. The heavy tax duty for cruise ship purchase, on one hand, explains why these cruise ships are registered abroad. The total of import duties and VAT add up to nearly 30%. There are no many destinations available to cruise tourists. For the moment, China's cruise industry still relies on basic services including port berthing and provisioning. Coastal tourism resources are not integrated with cruise tours and new patterns of development are needed for port economy. At the moment, the main development pattern of China's international cruise tourism is that Chinese tourists take luxury cruise ships owned by foreign cruise lines to visit other countries. China, in spite of abundant ports and unique oriental tourism resources as well as world- famous tourist destinations, has no successful cruise destinations. New paths for developing cruise terminal economy are needed. At present, most cruise terminals consider cruise home-port as the objective of development, believing that the economic effects of a cruise home-port are much greater than a visiting, departing or calling port. However, according to statistics of ports in Japan and South Korea, the 1:10/1:8 economic benefits failed to show up, while huge economic benefits were generated by those visiting ports. Rational planning shall be made for all ports. The focus shall be on the economic benefit, the experience of tourists and the life of local residents, not the construction of a home port. Greater innovation is needed for port supervision. Lots of ports have introduced preferential policies including "144-h visa-free transit" and "15-day visa-free entry". To attract more foreign visitors and develop local cruise tourism, a longer period of stay may be granted and the entire visa-free system may be gradually introduced to cruise ports. Greater innovation is needed for cruise route approval. Local cruise ships need to operate on domestic routes. The further advance to upstream of the South China Sea route will be an important acting point to promote further development of local cruise tourism of China.

3.7 There Is Insufficient Innovation in Marketing and the Drawback of the "Chartered" Mode Starts to Show

Distribution channel plays a critical role in all commercial activities. The "chartered" mode, a unique mode of distribution, has brought many drawbacks including the price war, lowered quality of product and service, lowered satisfaction of tourists, upside down relationship between chartering price and price of retail route, impacted profit of the charter and the cruise line, disputes between the charter and the cruise line and frequent occupancy of ships by tourists. In the western market where the cruise economy is relatively mature, the primary mode of distribution is "retail", i.e. the cruise line directly sells tickets to customers, or the agency factor of ship tickets by travel distributors who utilize their wide distribution network for ticket selling and then collect proper commissions from cruise lines based on mutual agreements. In this mode, the cruise line can directly find out real demands of customers and obtain real market feedbacks.

The reason that the retail mode popular in the western developed market has not grown rapidly in China is because (1) the trend of oligopoly competition has long been formed in China's tourism market and emerging tourist modes must rely on powerful distribution channels. As most cruise lines are overseas homed, the development of new markets must depend on local distributors. Foreign cruise lines are willing to transfer risks of the cruise market to powerful distributors; (2) cruise lines have not established powerful brands and distribution channels. With the rapid development of the Internet and the mobile Internet and fast growth of the APP and platform economy, it usually takes huge financial resources, effort and manpower to obtain a large quantity of customers. Platform economy is a mode of economy the burns money the most; (3) in European and American countries, tourists have better idea about cruise lines and are familiar with various cruise products. Cruise lines make great efforts in publicizing their products so that many tourists choose to purchase products from their own channels; and (4) charterers want to get the larger piece of cake in the new tourism product market, obtain more resources from cruise lines, increase the market influences and allocate resources in a more direct manner in the "chartered" mode.

4 Development Trend of China's Cruise Industry in 2018

China's cruise market is experiencing transformation and becomes mature in rapid growth and in the pain. In the future, major changes are expected in product structure, product quality, service quality, development of distribution channels, brand building, the cognition of tourists for cruise tours, the acceptance of tourists for high-quality paid services, civilization level in tours, etc., a glamorous aspect of the construction of a well-off society. Cruise lines place high hopes on the Chinese market. Some foreign cruise lines are placing larger, more luxurious and newer cruise ships in China that may lead to more fierce competition in the Chinese market. Local cruise lines will also purchase more cruise ships to form larger fleets, aiming to become the main force of the market. International cruise lines are not only turning China into the home and visiting port, but intend to invite more Chinese tourists to their overseas routes. Local companies will also explore more overseas products and even March into the special cruise market such as adventure cruise tours.

4.1 The Development of the Country's Cruise Tourism Industry Will Focus More on the Globalism and Coordination

The development of the cruise tourism industry shall be based on overall national planning including the construction of international cruise terminals, the provision of international cruise services and the regional linkage of different markets so as to provide an overall guidance for developing the cruise tourism industry nationwide. As to the construction of cruise terminals, many coastal cities of China are building international cruise terminals or reconstructing existing cargo wharfs to guarantee the capability of berthing cruise ships. International cruise terminals have been successively built in Tianjin and Dalian of North China, Qingdao, Yantai, Shanghai and Zhoushan of East China, Xiamen and Fuzhou of Southeast China, Guangzhou, Shenzhen, Sanya and Haikou of South China. A new cruise terminal is currently under construction in Beihai, Guangxi of Southwest China. Most cruise terminals are oriented as international cruise home ports, with related services carried out proactively, leading to an obvious trend of homogenization. Despite national planning documents prepared for the construction of international cruise terminals, the construction and planning of international cruise ports in current phase are mostly completed by local government and enterprise and not included into the national overall planning. As a result, lots of terminals receive ships less than scheduled and numerous facilities are in idle state, leading to poor profitability and overall economic efficiency.

In 2018, China's cruise industry will focus on the overall planning to achieve clustered industrial development, highlighted regional development and the complimentarily of regional development, so as to realize coordinated development of home port and visiting port to promote overall advance of China's cruise industry. With respect route design, supplementary development will be the priority.

4.2 The Development of the Country's Cruise Tourism Industry Will Focus More on the Stability and Sustainability

China's cruise tourism market is transforming from "rapid" to "steady" development, reflected in the steady growth in the placing of new types of cruise ships, the number of ships received at international cruise terminals and the scale of inbound and outbound cruise tourists. During the first half of 2016, both ships received at cruise terminals and inbound and outbound cruise tourists received at terminals grew at a rate over 100%. However, in the first half of 2017, the growth rate decreased abruptly to around 30%. This is directly related to the significant increase in the base number of the cruise market, also a mark that the cruise market is growing into maturity. In 2016, the scale of outbound tourists reached 122 million passengers, comfortable perched at the top of the global origins of cruise tourists. On this basis, China became the second largest cruise market of the world in 2016 with a total scale of cruise tourists of 2.10 million passengers.

The potential of the cruise market depends on the market penetration. As long as the market penetration maintains a high growth rate, the cruise market will see greater potentials in the future. In the category of economics, market penetration means the comparison of current consumption demand with potential market consumption demand. However, certain preconditions must be satisfied for the establishment of market demand. In cruise tourism consumption, there must be a certain group of tourists who have special preferences for cruise tours, a certain cost for cruise tourism and the power of consumption of a tourist. These factors will finally decide whether the consumption demand can be realized. When calculating penetration of the cruise tourism market, the total tourists in the cruise tourism market were compared with the total population of a country or region. In 2016, the scale of China's cruise tourism market reached 2.10 million passengers. Considering a total population of 1.3 billion, the market penetration will be 0.16%, which is far away from that of Europe and North America where the cruise tourism is highly developed. Such figure shows that there are great potentials to be met by China's cruise tourism market. However, regarding the outbound tourism market, among the 122 million passengers in 2016, there were 83.95 million passengers visiting Hong Kong, Macao and Taiwan, accounting for 68.8% in total outbound tourists of China. That is, there were only about 28 million passengers who had experienced the real overseas tours. Taking the outbound tourists scale as 28 million passengers, the market penetration is 7.5%, which is already at a high level. Of regions showing the strongest willingness for outbound tours, Beijing, Shanghai and Guangzhou rank the first three, consistent with the ranking of the three main origins of cruise tourists of China and showing high market penetrations of the home-port cruise market.

In sum, China's cruise tourism market will develop in a more resilient and steady manner, able to adapt to changes of the external environment and transforming from quantitative to qualitative promotion. Cruise lines will pay more attention to promoting quality of cruise products and services and place more talents and efforts into the cruise market.

4.3 The Design for China's Cruise Tourism Products Will Be More Characteristic and Differentiated

The vigorous advocacy of international cruise lines and travel agencies enable Chinese tourists to know more about cruise tourism, leading to a rise of rigid demand for cruise tours. Chinese tourists have gradually turned their attentions from product price to product quality and features. In the Chinese market, differentiation is a key to gaining advantages out of a great variety of cruise tourism products, and more and more companies have begun to differentiate their products. Hurtigruten aims at adventurers and is the only cruise line launching the cruise route to North and South poles, who no longer merely emphasizes the concept that "the cruise ship itself is the destination", but also attaches great importance to off-board sceneries and makes the cruise ship a base for adventures. Costa Cruises exclusively launches the Cruising Italian Style. SkySea Holding International Ltd., a local cruise line, differentiates their products by highlighting the Chinese style and higher quality of services. In 2017, the concept of "First Class at Sea" was, for the first time, introduced to the "Norwegian Joy" of Norwegian Cruise Line which was customized for the Chinese market. Royal Caribbean Cruises Ltd. vigorously highlights the concept of "Maritime Mansion" in 2017.

Currently, many cruise lines have introduced the concept of "tailored for China" and placed their latest ships into the Chinese market, showing their prior considerations of the Chinese market. Some companies even included elements favored by Chinese into their products during the design, research, development and manufacture processes. Chinese elements are also observed in catering, recreational and other aspects. "Tailored for China" is a dynamic process that combines western cruise culture with the Chinese culture and aims to provide comfort and convenience for tourists during their tours and give them more impressive feelings. The "Majestic Princess" placed by Princess Cruises into the Chinese market in July 2017 is called "a luxury and master-grade cruise ship customized for Chinese guests", which aims to provide master-grade products and services for China's cruise market. Other cruise ships "tailored for China" include the "Ovation of the Seas" owned by Royal Caribbean Cruises Ltd., the "Genting Dream" owned by Dream Cruises, the "Norwegian Joy" owned by Norwegian Cruise Line. Some other cruise lines have accelerated the "localization" process, including the "MSC Lirica" owned by MSC Cruises and the "Skysea Golden Era" owned by SkySea Holding International Ltd.

Each brand has its own features and advantages and market sections. Different cruise ships are placed in different market sections. However, however the differentiation is, it is always based on concrete services and products and thus to establish featured and highly recognizable cruise brands in the market with high brand awareness, as tourists always care about products, services and the final brand itself.

5 Countermeasures for Achieving Steady and Sustainable Development of China's Cruise Industry

5.1 Enhance Publicity and Advocacy of Cruise Tours and Establish Reputation of Cruise Tourism

Although having developed over a decade, cruise tourism is still an emerging travel mode in China that is characterized by maritime travel destination + outbound touring at the visiting port. Even if China's cruise market has risen to the second of the world in terms of scale, the number is merely 2.10 million, meaning a still small-crowed travel mode in China's enormous tourism market. Some people know nothing about this travel mode. In the past few years, the cruise culture was promoted at a slow pace. Most tourists focused on the onshore touring at the destination and attached too much importance to cost performance of cruise products instead of the essential attribute of leisure of the ship itself, who always considered luxury ships as a kind of relatively cost-saving and comfortable maritime means of traffic. It is not easy to convert the cognition of Chinese cruise tourists for cruise tours. In the history of China's tourism, people always took natural and historical and cultural sceneries as the destination of a trip, especially scenic areas rated above Class-A. Cruise tourism, however, is a product of the western culture, which is hard for Chinese tourism to deeply understand and take it granted that the ship itself is the destination. Therefore, more efforts shall be made to expand such cognition.

Both cruise lines and travel agencies shall undertake the liability to promote cruise culture. Once deeply understanding cruise tours, tourists will obtain higher "cost performance" from these tours. At present, the cruise culture in China is immature. The market cognition for cruise tours and cruise brands is at a low level. The urgent need is to vigorously promote and advocate cruise tours and convey information about cruise tours to the public. To expand the cruise market, it is essential to improve the universality of cruise tourism and enable more people to know about, understand and be fond of this mode of outbound touring, which is the key to enlarge China's cruise market.

More advocacy efforts are needed in densely populated places including commercial plazas, pedestrian zones, leisure plazas, subways and subway stations, expressway advertisement boards, high-speed rail stations, main tourist sits, etc. Understanding is the precondition for the willingness to buy. The government may vigorously promote local cruise ships by publishing more news reports on cruise tours, setting the cruise tourism festival, the cruise tourism year and actively promote the interaction of cruise ships in coastal areas.

5.2 Improve Quality of Cruise Products and Enhance the Experience in Cruise Tours

At present, China's cruise ship products have developed from strong combination of products abroad into a centralized package of tourism products. In the future the more free combination of cruise products will be introduced to attract more tourists with characteristics and diversity. Efforts will be made to enhance the added value of products and enhance visitors' experience. Both cruise lines and travel agencies shall undertake the liability to promote cruise culture. Once deeply understanding cruise tours, tourists will obtain higher "cost performance" from these tours.

Cruise tourism is a high-end mode of travel, but the concept of "high-end" can not only be reflected in the high price, but specifically embodied in the high-quality products, high-quality services and the whole-process high-quality experiences: classic, grandeur and whole-hearted service. Tourism itself is a way for tourists to relax, to escape the reality of pressure and enjoy the way of life. In the whole process of cruise tourism participated by tourists, comfortable and convenient service must be actively provided for tourists.

Before buying the product, tourists understand the process of cruise tour. Service providers shall be patient and patiently communicate with the visitors, standing from the point of view of the visitors and satisfying their demand for detailed explanation, to enable them to feel high-standard consulting services. This will make the tourists desire to buy, and then realize the purchase behavior.

Timely communication shall be made to tourists after they buy tourism products to find out their needs and concerns, and ensure high-quality services are available to them. Traffic is always a major concern for tourists, especially those from inland areas who need a long-distance trip. Convenient services shall be provided to satisfy traffic needs of tourists. Suggestions may be offered to help tourists to set their schedules. The cruise service provider may also communicate with other traffic sectors to provide whole of half-distance traffic services to ensure tourists arrive at the ship fast and conveniently. The objective is not only to provide high-quality services for tourists, but increase revenue of the service provider.

Good services shall be available even before tourists arrive at the terminal, at least guarantee the basic comfort of tourists. The service provider needs to seek for policy support to optimize the embarking process and make proper arrangement for diversion so that tourists can get on board in time. On the ship, tourists shall be provided with high-end cruise products and services, which is currently in good practice but requires more on-board projects. For onshore touring, new destinations shall be explored to deeply impress and entertain tourists. At the end of a trip, feedbacks from tourists shall be collected and inputted into the database. The big data technology shall be applied to ensure accurate marketing in the future.

5.3 Proactively Innovate the Distribution Mode and Eliminate Drawbacks of the "Chartered" Mode

The "chartered" mode has been gradually given up by the market mainly due to the asymmetrical return-risk relation and increasing market risks. Cruise products are sold by cabin price, which is obviously featured in value fugitiveness and timeliness. The greatest concern of a charterer is to distribute cruise ship cabins within a short time. However, cruise tourism is fragile and subject to external uncertainties such as international relations. Recently, there is an obvious large- sized trend of cruise ships entering into the Chinese market. Large quantities of cabins can hardly be sold out merely depending on self-owned distribution channels. More channels are needed to distribute chartered tickets and thus charterers have to depend in the industry and even include travel agencies not qualified for outbound business into their channel systems. The game between charterers and cruise lines transits to that with multiple parties, threatening the interest of charterers. Therefore, the fundamental reason that the "chartered" mode is questioned is that the income of charterers is reduced, the overall market price declines and the advantageous position of cruise lines is threatened leading to lowered revenue. In the past decade, the "chartered" mode contributed greatly to the development of China's cruise market, but the year of 2017 calls for reform on the distribution mode of the market. Although travel agencies enhanced their efforts in promoting the "chartered" mode at the beginning of the year, with the operational stress increased in the cruise market, they began to gradually cut the "chartered" mode and replace it with cabin sectionalization, half-charter, retail, etc., which can involve more distributors into the distribution process so as to effectively lower and share risks of cruise lines and travel agencies.

The travel agency-based distribution mode is gradually transiting from B2B to B2C. More themed routes are designed to attract the C-end customers. More and more cruise lines start to independently distribute cruise tickets. Such mode of selling may be the mainstream in the future, provided that a powerful distribution network and a high customer flow are established. In the meantime, cruise ticket distribution channels are observed as "narrow and long" channels featured in layers of agency, leading to excessive agency in the chain, lower control of charters over subordinate agents and even the emerging industry of ticket reselling. The pricing system of the market and quality and reputation of cruised services are in danger. The first step is to build a blacklist of ticket resellers and disclose it to the public to crack down these resellers. Secondly, a proper rating system shall be established for distributors. Those having great distributing powers and high reputations are allowed with a greater quota. Thirdly, limitation shall be imposed on the quantity of cabin for sale, i.e. to specify the number of cabins sellable by each distributor and involve more distributors into the process to form a flattened distribution network.

Policies may be introduced to encourage cruise lines to set up their own distribution channels. At present, Royal Caribbean Cruises Ltd., etc. are attempting to set up their own distribution channels. However, it is extremely difficult to realize a sufficient amount of customers from independent distribution channels, especially in the Internet economy where platform giants have formed. Cruise lines have to obtain tourist information from travel agencies and thus cause certain market risks to the latter. Nevertheless, independent distribution channel is an inevitable trend for cruise lines that can obtain real demands of tourists in the Chinese market, improve the control over channel quotas, expand the market scale and ultimately promote the cultivation of cruise tourism culture. The cooperation with middle and high-end hotels, middle and high-end shopping centers, 4S stores and classic restaurants who have mature customer groups will effectively expand market resources via mutual recommendations between customers and resources integration.

5.4 Build a Favorable Industrial Soft Environment and Restructure the Cruise Industry Environment

Follow international cruise charging standards and perfect the charge mechanism of China's cruise terminals. Charges of Chinese cruise terminals at present include berthing fees, navigation fees, charges for tug's service, harbor dues, vessel agent fees, barge service fees, port construction dues as well as charges for garbage disposal, water supply, fuel supply and power supply. Tonnage tax is the main category of tax involved, including 2/3 stipulated fees. As to taxes and fees for international cruise terminals, many countries levy relatively low taxes and even cancel the tonnage tax.

The suggestion is that the Ministry of Transport adopt international charging standards and regulate pay items, gradually simplify the charge system and lower and even cancel the tonnage tax within 3 years; adjust the charging structure and separate basic charged services from competitive charged services; lower the standard for basic charged services and complete the charging structure for competitive charged services, and rationally classify charges into various levels. It is also suggested that terminal operation enterprises adopt the return management mode and gradually implement the flexible pricing mechanism for terminal charges, ensuring to realize coordinated and sustainable development of all interested parties by implementing different price levels depending on the demand characteristics of different cruise lines and price flexibility.

Carry forward the Policy Investigation and Research—Packaged Reforms on the Provisioning of Cruise Ships. The traditional provisioning for cargo ships can no longer satisfy the demand for the cruise industry, due to which the provisioning of cruise ships can not break the bottlenecks such as complexity of monitoring and supervision procedures, high cost for distribution and uncertainty in timeliness. Nowadays, many international cruise lines operating in china make supplies of their overseas purchased provisions including food, hotel supplies, etc. through visiting ports in Japan and South Korea instead of domestic ports. The category and quantity of supplies available for cruise lines to purchase domestically are very limited.

It is suggested to introduce special regulations in adaptation to characteristics of cruise ship provisioning and lower the taxes and fees for such provisioning; gradually allow provisioning enterprises to carry out global purchase as floating goods, which can be directly distributed to cruise ships from domestically ports; establish integrated bonded areas for cruise ship provisioning and vessel component import and extend the cruise industry chain; set aside fast channels for vehicles undertaking the provisioning of cruise ships and simplify the procedures for supervision; allow qualified cruise lines to independently carry out cruise ship provisioning business to enable healthy development of the industry.

5.5 Accelerate the Integration of Cruise Tourism into Regional Economy and Generate the Driven Effect

The formation of tourism economy is the necessary result of the joint participation of various market players. There will never be tourism economy without a considerable tourist scale, and the formation of tourism economy also requires the sustainability of such tourist volume and may suffer deadly damage once the tourist flow breaks. However, in today's cruise market of China, tourists are not considering the cruise ship as the destination of travel and a higher class of travel mode, thus their loyalty to cruise tours is relatively low, leading to strong demand price flexibility. Once significant change of price occurs, the market demand will suffer greatly. It is necessary to make cruise tours a maritime lifestyle for Chinese tourists, as in the western market. "Lifestyle" is a term naturally different from "travel mode". The former is formed during a long period of time and an integral part of daily life. To realize this objective, cruise lines and travel agencies shall work together. However, according to current statistics of advertisement launching and marketing, most promotions for cruise tours are undertaken by travel agencies, distributors and retailers. Cruise lines participate in this progress less than expected.

As the service provider of the cruise market, cruise lines need to ally with distributors to jointly publicize cruise culture and promote product and service innovation. The extension of the cruise industry chain and the development of cruise tourism economic zones are in many aspects similar to the idea of development from "cruise wharf" to "cruise terminal" and "cruise city". At present, China has numerous quality ports who claim to build cruise home ports. However, a cruise home port is a kind of heavy asset that is fixed and different from ships which can be easily transferred in case of changes in market focuses. Such swarmed construction of home ports cost a huge amount of financial resources and lead to major market risks. As a result, rational planning is essential to promote the joint construction of home ports, visiting ports and departure ports. Regions shall cooperate and coordinate with each other to develop local cruise economy with their own ports, port tourism resources and regional advantages. We also expected that, in the future, Miami and Barcelona will not be the only marker posts in global cruise terminal economy. The Chinese mode of development and path of realization will be borrowed by the cruise society.

5.6 Follow "The Belt and Road" Initiative and Realize Internationally Coordinated Development

Realize linked development of cruise terminals of countries along "the Belt and Road". Highlight regional advantages of China and the internationalized application of terminal standards. Cruise terminals play a critical role in the cruise industry. Over the past ten years, China's terminal infrastructures were being perfected and experiences were gained during port construction, leading to a stronger technical power and rich experiences in operation and management. The construction of international cruise home port will become a key aspect of "the Belt and Road"-base tourism. Countries and regions along "the Belt and Road" and the 21st Century Maritime Silk Road own abundant maritime resources which are ideal for developing the cruise industry. Cooperation will be continued with AIIB, New Development Bank, the World Bank and other multi-lateral development organizations to support Chinese enterprises to construct cruise terminal infrastructures in countries along "the Belt and Road" and promote regional advantages of China and the internationalized application of terminal standards.

Realize linked development of cruise systems of countries along "the Belt and Road". Shanghai is the leader of the cruise industry that can accelerate the output of intelligence and talents as well as operation and management experiences in China's cruise terminals to countries and regions along "the Belt and Road". Efforts shall be made to complete the educational and training system for the cruise industry and to form Chinese standards in the international cruise industry.

Innovate cruise tourism products along "the Belt and Road". 56 countries and regions along "the Belt and Road" will become the destination of outbound tours of Chinese citizens. Promote the "multi-port call" policy along "the Belt and Road" and diversify cruise routes of the world. Cruise tours are usually international and trans-regional and thus can integrate tourist resources, destinations and ports along "the Belt and Road". The promotion of the "multi-port call" policy is an important path of development and can diversity the world's cruise routes. Explore quality routes along "the Belt and Road" and regionally featured cruise routes. Accelerate the exploration of intercontinental and global long-distance routes with China as the home and visiting port.

Enhance cruise product marketing along "the Belt and Road". The tourism industry can enhance cultural exchange and civilization integration of people along "the Belt and Road". China has become an important destination for outbound tours of residents of many countries along "the Belt and Road". According to statistics of

CNTA, during the 13th Five Year Plan Period, China will attract a total of 85 million passengers from countries along "the Belt and Road" and realize a total tourism consumption of USD 110 billion approximately. Cruise tourism, as a high-end travel mode, has developed rapidly in China with China as the home port, but the inbound tourism has been in slow pace. The suggestion is to ally with countries and regions along "the Belt and Road" and build the brand of "Cruise Tourism—the Maritime Silk Road"; hold the "Year of Tourism of the 21st Century Maritime Silk Road and the International Cruise Tourist Festival of the Silk Road with countries along the line so as to increase the reputation and influence of the Chinese cruise tourism brand.

Promote exchange of cruise industry policies along "the Belt and Road". "The Belt and Road" initiative has provided a new development platform for the international cruise industry that effectively interconnect China with the western cruise market. The high openness of the industry needs the support from the think tank as well as international exchanges of the industry, academic circle, government agencies and industrial associations. Proactively establish relatively uniform international cruise industry development rules and standards with countries along "the Belt and Road". Cooperate with local regions to establish quality international cruise tourism exchange platforms and develop strategic cooperation partnerships among port cities.

Promote interconnection of cruise management standards along "the Belt and Road". Enhance the cooperation and exchange in cruise ship customs clearance, supervision, information, data and formalities. Simplify the process cost for cruise industry development. Further advance mutual recognition of law enforcement and sharing of data, information and standards. Proactively explore new patterns and paths for interconnection supervision. Promote facilitation of visa issuance of countries along "the Belt and Road" and try to obtain visa-free transit policies for Chinese cruise tourists from more countries. Promote traffic right opening, mutual recognition of certificates and tourism insurance.

Promote mutual aid in cruise talents cultivation along "the Belt and Road". Carry out international cruise tourism talents cultivation for "the Belt and Road" and special topic research. Professional talents are the primary resource to develop and expand the cruise industry. The cruise industry is a highly combined and international industry that requires highly internationalized talents who have global perspectives and views. The need for cultivating and attracting such talents is urgent. Lay a solid talent foundation for developing the cruise industry in countries along "the Belt and Road". It is suggested that higher educational institutions with accumulations in the cruise industry attract students from countries and regions along "the Belt and Road" to study the graduate and undergraduate courses of cruise under the uniform planning of the Ministry of Education and Shanghai Municipal Government.

Promote linked development of cruise industry along "the Belt and Road". It is suggested that Shanghai Municipality take the lead to form "the Belt and Road" International Cruise Industry Research Think Tank. Countries and regions along "the Belt and Road" will inevitable encounter many topics to be solved during the development of cruise industry, for which the professional ability and influence of such international cruise think tank can be exerted. Shanghai shall also take the lead to hold "the Belt and Road" Cruise Industry Conference, form "the Belt and Road" International Cruise Industry Association and hold seminars on "the Belt and Road" international cruise cultural system as well as set up new standards for the industry.

Chapter 2
Ten Hot Spots of China's Cruise Industry Development During 2016–2017

Hong Wang, Jianyong Shi, Younong Wang, Xingliang Ye, Ling Qiu, Ke Chen, Junqing Mei, Xie Xie, Mingyuan Wu, Ruihong Sun, Guodong Yan, Xia Li, Tian Hu, Taotao Li, ERbing Cai, Shuang Cai, Fengri Wang and Feng Yu

During 2016–2017, with the implementation of "the Belt and Road" national development strategy, the supply-side structural reform and guided by the 13th Five-Year Plan for the tourism industry of the country, the maritime tourism economy welcomes major development opportunities. Cruise tourism has gradually become an important part of China's tourism and an important engine that drives the globalization, internationalization and brand marketing of the tourism industry. In this process, some hot events aroused the unprecedented concerns of all circles from the public to researches, from domestic and foreign tourists to industrial experts.

To comprehensively view and accurately analyze these events, Shanghai International Cruise Business Institute conducted extensive online selection on June 2017 and finally chose ten hot events from over twenty candidates based on principles: events happening in the cruise industry in the past year that have aroused intensive responses and representative events that could give significant enlightenment to the development of China's cruise tourism. The final ten hot events were chosen based on the discussions of 142 industrial experts and scholars, especially with the discussions of professionals in this field in July 2017, of which the positive and negative influences were dug at depth to play the demonstrative role and bring along more enlightenment to the society as well as more momentum to the future of the industry.

Ten hot spots in China's Cruise Tourism Development during 2016–2017 (September 2016 to August 2017).

H. Wang (✉) · J. Shi · Y. Wang · X. Ye · L. Qiu · K. Chen · J. Mei · X. Xie · M. Wu · R. Sun
G. Yan · X. Li · T. Hu · T. Li · E. Cai · S. Cai · F. Wang · F. Yu
Shanghai International Cruise Business Institute, Shanghai, China
e-mail: wh@sues.edu.cn

© Social Sciences Academic Press and Springer Nature Singapore Pte Ltd. 2018
H. Wang (ed.), *Report on China's Cruise Industry*,
https://doi.org/10.1007/978-981-10-8165-1_2

1 Hot Spot 1 Cruise Industry Proactively Follows "The Belt and Road" Initiative and Develops a New Open and Cooperation Pattern for the Cruise Industry

1.1 Hot Event

The Belt and Road Forum for International Cooperation was held during May 14–15, 2017, in Beijing, the capital of the People's Republic of China, with the theme of "strengthening international cooperation, developing the Belt and Road and achieving win–win development", which, for another time, raised "the Belt and Road" initiative to a new high level. President Xi Jinping emphasized the spirit of peace and cooperation, openness and tolerance, mutual learning and borrowing as well as mutual winning and benefiting to ensure peace, prosperity, openness, innovation and civilization of countries and regions along "the Belt and Road". "The Belt and Road" initiative brings along great opportunities for the international tourism, which is playing a key role in spreading ideas, publicizing international cultures, assisting national diplomatic strategies and promoting peace-based win-win of the world and becoming a new bridge of international relationships.

To proactively integrate with the state's B&R proposal, Shanghai held the Symposium on "the Belt and Road" and Cruise Economy Development in June 2017, on which the grand objective of "building Shanghai an international cruise city" was proposed from five strategic aspects: building a world-famous cruise home port, building the most competitive cruise economy cluster in the Asia-Pacific region of B&R, developing an integrated demonstration area for China's cruise economy, developing core sections for the International Navigation Center and developing key nodes of international, intercontinental and global cruise routes of the 21st Century Maritime Silk Road. In this context, Shanghai will tightly grasp the strategic development opportunity and, with new development pattern, momentum and space, build Shanghai an international cruise city, cultivate the whole industry chain, innovate patterns for international cruise services and expand new spaces for the industry.

In July 2017, the "Majestic Princess"—the cruise ship customized to the Chinese market by Princess Cruises homed in Shanghai, arrived in the city on the route of the Maritime Silk Road. The ship departed Roma, Italy, on May 21, passed Athens (Greek), the origin of European culture and Santorini, Aqaba (Jordan)—the hub between Asia and Europe, Dubai (UAE)—the "Desert Oasis", Cochin (India)—the "Queen of the Arabian Sea", Colombo (Sri Lanka)—"The Eastern Crossing" and Singapore—"The Lion City", totally 13 ports, and finally arrived in the new home port Shanghai in July, marking the success of the "Tell China's Stories along the Maritime Silk road" promoted together with the Chinese People's Association for Friendship with Foreign Countries.

1.2 Comments on the Event

China has transformed from a country lacking the development of tourism to a large and powerful country of tourism. Driven by both the government and the market, the country's cruise tourism has grown from zero to a powerful player in the global market, with the steady progress of the state's tourism diplomatic policy. At a time when China's cruise market has entered into a key period of innovation-driven development and structural transformation and upgrading, "the Belt and Road" initiative further propels it into another prime period. "The Belt and Road", especially the 21st Century Maritime Silk Road offers a vast space for choosing routes of long-distance cruise tours. Diversified long-distance international cruise voyages become a critical option that supports the sustainable development of the market. The implementation of "the Belt and Road" initiative significantly enhances international and domestic coordination to further enlarge the international cruise market. "The Belt and Road" initiative has also provided a new development platform for the international cruise industry that effectively interconnect China with the western cruise market.

1.3 Important Enlightenment

The simplicity of cruise routes has always been a headache in China's cruise market. Routes to Japan and South Korea hold a proportion of over 95% of the total market, severely restricting the diversity of China's cruise products. The fragility of the market is even highlighted with the low level of cognition for cruise tours by Chinese tourists. "The Belt and Road", in all directions centered on China, gets through the channels to Europe, the Mediterranean and the Indian Ocean. In particular, the 21st Century Maritime Silk Road starts from Chinese ports, pass through the South China Sea and the Indian Ocean and arrive at Europe, getting through the channels from China to Europe, the Mediterranean and other regions where the cruise market develops the best, being an extension of China's international, intercontinental and interoceanic routes, making countries along "the Belt and Road" the largest destinations of China's cruise market and providing an enormous space for development of China's cruise industry. The suggestion is to enhance the international advocacy of China's cruise tours and destinations, encourage travel agencies and other organizations to set up overseas publicizing branches, create high-quality cruise routes themed Maritime Silk Road with enhanced international image and attraction of China among countries along "the Belt and Road", build national cruise tourism brands featured in Silk Road and maximize the promotion of inbound tourism of China.

2 Hot Spot 2 China's Cruise Market Has Risen to the Second of the World and Wusongkou International Cruise Terminal Has Joined the First Group

2.1 Hot Event

In 2016, total cruise tourists of China reached 2.28 million passengers, making the market the second largest of the world. In 2016, total cruise tourists of the world reached 24.70 million passengers, an increase of 6.5% from a year ago. According to credible statistics of CLIA, the total tourist volume of the global cruise market in 2017 will reach 25.80 million passengers, showing certain increase compared with the estimated 25.30 million passengers, indicating that the international cruise society is confident in the cruise market. Countries with an amount of cruise tourists exceeding 1 million include the U.S. (11.52 million), China (2.28 million), Germany (2.02 million), the U.K. (1.89 million) and Australia (1.29 million). This is the first time that China's cruise market tourist volume exceeded 2 million since 2006 when China entered into the era of home port. China is the largest cruise tourist market of Asia, holding a share of 47.4% in 2016, ranking the first. In 2016, Shanghai, Tianjin and Guangzhou ranked the top 3 by received volume in the cruise regional market. Shanghai received 509 ships, 50.4% of the national total, while Tianjin received 142 ships, accounting for 14.1% of the total number in the whole country. Guangzhou, as a follower of the market, showed strong growth momentum in 2016 and received 104 cruises throughout that year, accounting for 10.3% of the total number of the whole country, and received 325,900 passengers, accounting for 7.1% of the total number of the whole country. There are currently three centers in China's cruise market: Shanghai, the center of the East China market, Tianjin, the center of the North China market and Guangzhou, the center of the South China market. In which, Shanghai is still the most active player in the national market. In 2016, Shanghai Wusongkou International Cruise Terminal realized a tourist reception volume ranking the fourth in the world.

2.2 Comments on the Event

The year 2016 is a landmark in the history of China's cruise industry, a year when the 13th Five-Year Plan is implemented and the supply-side structural reform beings, and the first year of the "Prime Decade" of China's cruise industry. China's cruise industry experienced major reforms and innovations in this year, with the realized achievements exceeding far beyond the anticipation made years ago. This is attributed to the state's strong support, the full-strength promotion of industrial associations, the marketing activities of cruise lines, the rapid growth of tourism, the product innovation of travel service providers, the intelligent support from the academic society, the talent nurturing of higher educational institutions and the upgraded consumption in the tourism market. The continual growth of China's cruise market is closely related to the increasing demand for vacation spending and the prosperity of the tourism industry. In 2016, the outbound tourist volume of China reached 135.13 million, maintaining the world's largest origin of cruise tourists. In the cruise outbound tourism market, China is the second largest cruise tourists' provider of the world.

2.3 Important Enlightenment

China, as the largest tourist origin and tourism spending country, has witnessed rapid growth of its cruise market and increased status in the global cruise economy. China has had 18 home-port cruise ships and its international share has increased by twenty times in a decade, from 0.5% in 2006 to 9.6% in 2016. The transport capacity grows steadily. The requirement is to realize rational layout of cruise ship fleets and diversify cruise routes. China needs to upgrade brand operations of cruise products and introduce multi-level and diversified brand building strategies based on demands of the Chinese market. Innovated driving-force is the source of improving attraction of cruise products. Cruise products needs to be versatile to realize "cruise+". As numerous international cruise lines are busy contending for the Chinese market, keywords including "localized cruise ship", "local cruise line", "local cruise tourism", "made in China cruise ship", "off-shore cruise tour" have been popular topics in the cruise society, and the international influence and market share of China's local cruise ships will rise gradually. With the international cruise economy entering into the "China Era", the country needs to deepen its reform and innovation especially in cruise tourism international cooperation, customs clearances procedures, innovation of cruise products and promotion of tourist spending so as to create favorable conditions for continual healthy development and reform of China's cruise industry.

3 Hot Spot 3 The First Industrial Fund for Cruise Industry Is Founded and China Force Joins the Building of Luxury Cruise

3.1 Hot Event

The localization strategy for cruise industry has always been a highlight in China. On December 28, 2016, the first domestic cruise industrial fund was founded, with the initial scale of 30 billion yuan, registered in Baoshan District, Shanghai. The fund is led by CSSC and jointly founded by six participants on the principle of "equal shareholder-right, shared interest and shared risk". The fund is management by CSSC Cruise Fund Management Co., Ltd. (hereinafter the "Management Company"), which is headquartered in Baoshan Industrial Zone and formed as a joint venture, with a registered capital of 300 million yuan. Shareholders include the 6 investors of the fund and Shanghai Wusongkou International Cruise Terminal, under which is set up special committees and professional work teams to invest in specific projects in a marketized way.

On September 23, 2016, on the 11th China Cruise Shipping Conference and International Cruise Expo Working Conference (CCS11), CSSC, China Investment Corporation, Carnival Corporation & PLC, Fincantieri, CSSC Cruise Technology Development Co., Ltd. and Shanghai Waigaoqiao Shipbuilding Co., Ltd. jointly signed the letter of intent for building 2+2 133,350 t (total tonnage) luxury cruise ships. The aforesaid cruise ships will be jointly built by a joint venture of CSSC and Fincantieri and are expected to be launched and put into the Chinese market in 2023 and 2024. As the first international cruise industrial zone, Shanghai CSSC International Cruise Industry Park opened in Shanghai, aiming at providing professional ship building, operation and supportive services for China's cruise industry so as to realize the optimization and upgrading of the whole industry chain, integrate with the Made in China 2025 strategy, accelerate the development of strategic emerging industries including smart manufacture, etc., and level up regional industrial structure, industrial energy level and industrial layout.

3.2 Comments on the Event

3.2.1 Follow the Concept of "The Belt and Road" and Promote the Development of the Cruise Industry

In 2015, NDRC, the Ministry of Foreign Affairs and the Ministry of Commerce jointly issued the Vision and Actions on Jointly Building Silk Road Economic Belt and 21st Century Maritime Silk Road, suggesting enhancing tourism cooperation, expand tourism scale and promote cruise tourism cooperation of 21st Century Maritime Silk Road. The building of luxury cruise ships is still one of the difficulties not yet tackled by China's shipbuilding industry, a dream of our shipbuilders and even the manufacturing industry. Currently, most luxury ship builders are in Europe. Fincantieri (Italy), Meyer Neptun Group (Germany) and STX France hold over 90% market shares of the global market for middle and large-sized cruise ships. The signage of the industrial fund is a milestone in history of China's shipbuilding industry that can accelerate localized design and building of cruise ships and promote the realization of the strategic objective of designing and building large-sized cruise ships in Made in China 2025. The concept of "the Belt and Road" facilitates the upgrading and innovation of "made in China", realizes the upgrading of the whole industry chain and is highly coordinate with the strategic objective of "the Belt and Road".

3.2.2 Introduce Flowing "Financial" Water to Irrigate the Cruise Industry and Set Up the Example of "Finance Supports Entity"

The Cruise Industrial Fund combines all interested parties as a whole. Its foundation marks the deepened integration of industry and finance in the history of China's shipbuilding industry, also a model of the "finance supports entity" economic mode, which facilitates to prevent and control investment risks, accelerate the transformation and upgrading of high-end equipment manufacture as well as promote the supply-side reform of the shipbuilding industry and health development of cruise economy. The fund works like the flowing water to irrigate the entire cruise industry chain, meaning the support for the budding industry by financial means. The fund will contribute to sustainable growth of the cruise industry and is a useful exploration of the "finance supports entity" economic mode and for the development of high-end equipment manufacture.

3.3 Important Enlightenment

3.3.1 China's Cruise Industry Develops Rapidly and Lots of Good Opportunities Come in the Cruise Ship Building Field

China's cruise market is the fastest growing section of the global market. In the past decade, cruise tourists of China increased by 320 times. It is estimated that by 2030, the number will reach nearly 20 million passengers, creating a huge demand for the cruise industry. The manufacturing centers of large-sized cruise ships are still in Europe, with an annual productivity of 7–8 ships. However, the demand for cruise ships each year is 15 ships, about twice the supply. The demand-supply balance is broken. The waiting list has already extended to 2025. As the cruise tourism market of China is in the ascendant, the imbalance in supply and demand of cruise ships creates even eager demands for building local cruise ships. In a context that supportive services need upgrading and the state encourages the introduction of high-end equipment for transformation, carries out the supply-side reform and upgrades tourism consumption, more and more opportunities rise for the investment in cruised ship operation, building of local cruise ships and local supportive services focusing on incremental demands of the domestic cruise industry. There is a prospective market. Faced with major opportunities brought along by rapid development of the Chinese and Asia-Pacific markets and significant increment of cruise ship demands, China needs to speed up the pace of localized building of large-sized cruise ships.

3.3.2 Develop Good Relations with Neighbors and Realize Localization of Cruise Ship Building Under Joint Efforts

Made in China 2025 clearly suggests to make breakthroughs in manufacturing technologies of luxury cruise ships. According to Implementation Opinions on Promoting the Tourism Equipment Manufacturing Industry jointly issued by MIIT, NDRC, etc., the urgent need is to accelerate the design and manufacture of cruise ships, promote the technical and commercial cooperation between powerful domestic shipbuilders with domestic cruise ship designing and manufacturing enterprises to make breakthroughs in both design and manufacture. The cruise industry is in nature capital-intensive and highly monopolistic. The foundation of the fund and the active participation of interested parties guarantees manpower and materials for developing local cruise industry and will accelerate the growth of a full industry chain. The monopoly in operation and the high technicality in manufacture have decided that China needs to closely cooperate with domestic cruise lines and shipbuilding companies to make breakthroughs in operation and building. In the future, domestic enterprises shall keep and enhance international cooperation and make good international relationships based on preferential policies to realize localization of ship building under joint efforts, thus to provide prime opportunities for structural optimization, supply-side reform, steady growth and transformation of China's cruise industry.

4 Hot Spot 4 Shanghai Takes the Lead to Innovate the Mode of Provisioning for Cruise Ships and Makes Great Effort to Build the Asia-Pacific Center for Global Cruise Ship Provisioning

4.1 Hot Event

Shanghai Wusongkou International Cruise Terminal has innovated the mode of provisioning for cruise ships "quarantine without inspection" to improve the provisioning efficiency of cruise ships. From the beginning of June to the end of July, the port totally completed container provisioning of totally 28 batches for the "Norwegian Joy". The "one opening" unpacking policy was implemented and the supervision flow was streamlined. Jointly witnessed by the ship owner, the port operator, the customs and the inspection and quarantine authority, the supplies for cruise ships from Spain and Italy were successfully loaded onto the ship at Qingdao Cruise Terminal on June 8, 2017, marking the success of Qingdao Cruise Terminal in operating provisioning of imported goods for cruise ships. In the provisioning field, ship owners purchase supplies from worldwide and transport them to bonded warehouses of a major port in containers. Containers are transported to the cruise terminal on the same day when the ship berths for unpacking and ship loading.

Loading of contained goods will be under the supervision of local customs, inspection and quarantine authority and the terminal operator to create a "through train" for loading goods from imported containers onto cruise ships. On June 7, 2017, Xiamen International Cruise Terminal implemented a new monitoring mode: directly delivering imported goods to cruise ships under bond, meaning that food supplies, after imported, are directly delivered to a specific cruise ship quarantined but not inspected provided that relevant regulatory requirements are satisfied, such that the inspection and quarantine procedures for bonded ship supplies/food are simplified, and the advantage of the bonded warehouse within the specific area supervised by the customs is full taken to realize the "one-stop" operation. With supplies loaded onboard through "fast channels", the efficiency of the provisioning for cruise ships is significantly improved and the cost for customs clearance is significantly reduced.

4.2 Comments on the Event

Cruise lines are continually increase the transport capacity in the Chinese market. More and more cruise ships operate homed at Chinese ports. However, the quantity of local purchase is limited and it is not convenient to operate ship provisioning from abroad, which adversely affect operations of ships and restricts the arrangement of routes. Completing the provisioning for cruise ships at the home port can ensure normal operations of the ships, facilitate route planning and also help cruise lines to generally arrange logistics and lower the cost. Provisioning plays an important role in ship operations. With respect to current policies towards the provisioning for cruise ships, it has not been officially included into the trading customs declaration management system and still follows the procedure of traditional cargo ships. There are complex formalities during customs clearance, tax refund and international transfer of supplies for cruise ships, which must be first stored in bonded warehouses before loaded onboard. High taxes and fees lead to the lack of competitiveness in terms of price. It takes a long time and many monitoring procedures to finish the inspection and quarantine procedures for transit goods. As a result, foreign cruise ships choose to take supplies of daily goods, food and consumables in Japan and South Korea. Only some fruits and vegetables are supplied domestically. The 17% VAT is an important influencing factor.

4.3 Important Enlightenment

Cruise lines locally purchase ship supplies and suppliers can not enjoy the tax refund. This leads to the lack of price competitiveness of goods directly purchased from inside China and those purchased through "global procurement and distribution", severely damaging the enthusiasm of cruise lines in local purchase.

Implement a new monitoring mode: directly delivering imported goods to cruise ships under bond. An integrated bonded zone and warehouse may be set up for cruise ships, with special approval procedures and supervision modes adopted to streamline the formalities and advance the provisioning industry. The current approval on cruise ship supplies providers is strict. There are only 3 provisioning operators. The beneficiary of tax refund is also the provider. The suggestion is to allow cruise lines to independently carry out provisioning for ships to improve their income-earning business, especial Chinese-funded cruise lines. The policies on the free trade zone (FTZ) may be comprehensively extended to cruise terminal zones, or to build new FTZs in cruise terminals zones and set up integrated bonded zones to satisfy the demands for ship supplies and ship component maintenance and repair, providing an ideal environment for extending the cruise industry chain.

5 Hot Spot 5 New Voices Rise in International and Domestic Cruise Conferences and Forums, Exploring a New Chinese Model of International Cruise Industry

5.1 Hot Event

The cruise industry of China has been developing rapidly, with its market size having risen to the second in the world. And the cruise industry of China has been making greater and greater contributions to national economy as well as global cruise industry, which has attracted extensive attention from all sectors of society. There were brilliant and diversified international and domestic conferences about cruise industry in 2016.

For example, the Global Cruise Industry Conference and the Seatrade Cruise Asia Pacific were held in Baoshan District, Shanghai, in March and October, 2016, respectively, where domestic and foreign experts from governments, enterprises, and research institutions conducted heated discussion about the trend of global cruise industry, offered advice and suggestions for the new round of development of Chinese cruise industry, and shared opinions on the future of cruise industry. In July 2016, the China Cruise Shipping Conference themed at "The Development and Cultivation of Asian Cruise Tourism Destinations" was held in Tianjin, during which a cruise procurement forum was set up for the first time and various experts discussed about the access of Chinese enterprises and products to the international cruise procurement system. In May 2016, the Northeast Asia Cruise Industry.

International Cooperation Forum and the 4th China (Qingdao) International Cruise Summit was held in Qingdao, during which such topics as the win-win cooperation model of Northeast Asia cruise industry, and the latest future market strategy and operation deployment of cruise lines in Asia and in China were discussed.

5.2 Comments on the Event

The cruise industry of China is having greater influence on global cruise industry. The international and domestic cruise conferences held in various places in 2016 attracted more and more attention from the world, with ever increasing influence, which has reflected the rising status of Chinese cruise industry in the world. In fact, the market of Chinese cruise industry rose to the second in the world in terms of size in 2016, being one of the fastest-growing cruise markets worldwide. The cruise industry of China has become an important engine for the development of global cruise industry. It is because the influence of Chinese cruise industry has been increasing that the influence of relevant international and domestic conferences is also increasing.

Relevant organizations are further focusing on the cruise industry of China. The international and domestic cruise conferences held in Shanghai, Tianjin, Qingdao, Xiamen, and other major port cities in 2016 gathered experts of major cruise market countries. They were from governments, industry associations, international cruise lines, competent organizations of international cruise cities, cruise ports, travel agencies, and other cruise-related organizations. There are two main reasons. On the one hand, the existing achievements of Chinese cruise industry have drawn global attention; on the other hand, the Chinese cruise market has great potential, but it also has a different environment, situation of supply and demand, competition pattern, etc. than those of other countries, with unique characteristics, so only by further focusing on and understanding Chinese cruise industry can the market be better exploited.

The development of Chinese cruise industry benefits from strong intellectual support. The Chinese cruise industry started late, requiring intelligence to develop. Strong intellectual support empowers Chinese cruise industry to maintain continuous, rapid and healthy development and to compete with the European and American markets. At the international and domestic conferences about Chinese cruise industry held at various places in 2016, based on different themes, a variety of topics about the development of Chinese cruise industry were discussed and the Chinese rules in international cruise industry were explored. The collision of ideas between relevant organizations and experts as well as lessons from cutting-edge research results provide solid intellectual support for the development of Chinese cruise industry.

5.3 Important Enlightenment

Firstly, international cooperation should be further enhanced to develop Chinese cruise industry. The international and domestic conferences about cruise industry have provided an international communication and cooperation platform. On the one hand, this platform brings the voice of Chinese cruise industry to the world and increases its international influence; on the other hand, the platform expands international communication and cooperation and draws us closer to the global cruise market. In fact, cruise industry has a distinct international nature. For example, one international cruise route usually consists of a home port and visiting ports distributed in different countries. Hence, only by further enhancing international cooperation for win-win results can the international competitiveness and influence of Chinese cruise industry be increased.

Secondly, the synergistic effect between industries is required to develop Chinese cruise industry. The above-mentioned conferences gathered experts from not only the cruise industry, but also other relevant industries. Cruise has strong industry relevance, and integrative development with other industries is the main approach to extending the value chain of cruise industry and increasing its contribution to national economy. To give play to the synergistic effect between industries can not only increase the energy level of cruise industry, but also expand the spread effect of cruise industry by means of industry relevance and increase its influence on primary, secondary, and tertiary industries as well as its contribution to regional economy. For example, the multiplier effect on employment will be shown.

Thirdly, further intellectual support is needed to develop Chinese cruise industry. The cruise-related international and domestic conferences are also the intellectual platforms for industrial development. The market environment and supply and demand structure facing Chinese cruise industry have their own characteristics. For example, in terms of supply, Chinese economy has entered the New Normal stage where the main tasks are stabilizing the growth, restructuring, and structural reform of the supply side. Under this background, Chinese cruise industry should make use of its advantage in stimulating domestic demand and grasp the opportunity to develop. In terms of demand, Chinese cruise tourists are different from foreign tourists in cultural background, consumer preference, behavior characteristics, etc., so further research on the Chinese cruise market is needed and the development of Chinese cruise industry still requires continuous intellectual support.

6 Hot Spot 6 with Increasing Attraction of Chinese Cruise Market, Famous International Cruise Lines Compete to Capture the Chinese Market

6.1 Hot Event

According to the latest statistics, the number of Chinese cruise tourists has exceeded that of Germany and ranked the second in the world. In view of this, cruise lines are swarming into the Chinese market, including Norwegian Joy, Majestic Princess, and Genting Dream.

On June 28, 2017, Norwegian Joy conducted its maiden voyage in Shanghai. This cruise ship has a total tonnage of 167,700 tons and a passenger capacity of 3883, known as "First Class at Sea". From the perspective of insiders, this is a trial made by Norwegian Joy in China, which tends to enter the Chinese cruise market with big ships. According to Frank Del Rio, President and CEO of Norwegian Cruise Line, Norwegian Joy is a large cruise ship tailored for China.

On July 9–10, 2017, Carnival Corporation held the China maiden voyage celebration for Majestic Princes in Shanghai. Majestic Princess is the first master grade deluxe cruise ship tailored for the Chinese market by the "Global Travel Master", Princess Cruises. Majestic Princess is 330 m long, with a tonnage of over 143,000 tons, 19 decks and 1780 cabins. Each of the cabins is provided with beds customized for Chinese guests.

In November 2016, the first cruise ship of Dream Cruises, "Genting Dream", had its maiden voyage from Guangzhou. The "Genting Dream" has a displacement of 150,000 tons and a capacity of 3400 passengers, with 2000 crew members, each of whom will serve 1.7 passengers. According to the responsible person of Dream Cruises, the "Genting Dream" will operate with Guangzhou (Nansha Port), Hong Kong, and Sanya as its home ports. Tourists may choose the model of "plane + cruise", boarding their ships at one home port and getting off at another home port to go on their journey. The flexible joint home ports model enables the tourists to fit the comfortable cruise journey into their schedule to realize in-depth travel at their destinations.

Under the increasing investment in China by global cruise tycoons lies the Chinese cruise market drastically growing in recent years. As estimated by the Ministry of Transport, the number of Chinese cruise tourists will reach 4,500,000 in 2020, becoming the most energetic and the largest cruise market in Asia-Pacific region. Facing the Chinese market with huge growth potential, Star Cruises made their big move on July 6, 2017. The "SuperStar Virgo" carried nearly 2000 Chinese tourists onto the "Golden Route of Fascinating Japan" from Shanghai. And hundreds of Japanese and overseas tourists will also board their ships in Tokyo and Osaka for a colorful summer journey. The 8-day-and-7-night golden cruise route is a rare long-distance product in current Chinese market. It is an elaborately designed unique route from the classical Osaka to the modern Tokyo, the majestic Mount Fuji, and the natural Kagoshima. Tourists only need to pack their luggage for once. Once they board the ship, they'll save their fatigue from transfers and long travel while visiting the four "must-sees" in Japan in a leisurely way. This voyage season will last till 2018, and the tourists may have the chance to appreciate the breathtaking sugarbush sceneries of Japan.

6.2 *Comments on the Event*

In recent years, the cruise ships of major cruise lines designed for China have shown the features of large capacity and new design, which has attracted a great number of domestic tourists. And the tickets are even hard to get during peak seasons. According to several cruise line senior managers, there are lots of opportunities in the Chinese market, and the competition between different cruise brands will benefit the industry. For the cruise industry, the primary task is to increase the awareness and understanding of Chinese tourists of cruise tourism. With the explosive development of cruise tourism in Asia, there are more and more diversified cruise routes and themes. The domestic cruise tourism will soon usher in its 2.0 Era with product differentiation. It is not a war of price, but a battle of IP.

The brand-new 8-day-and-7-night "SuperStar Virgo" route is a new model of long distance, fixed route and mutual home port voyage in Chinese cruise market, bringing a sunny journey for the Chinese tourists. This unprecedented route is an important attempt to build cruise routes that are in line with international practice and to provide diversified cruise products for Chinese tourists, and it will surely be a model of cruise product differentiation in Shanghai. As for the newly open route, Ms. Wang Hong, the party secretary of Baoshan District Committee, said, "As the pioneer of Asia-Pacific cruise industry, Star Cruises has offered the first experience of pleasant cruise tourism for many Asian tourists. Now it has returned to the home port of Shanghai and offered high-quality in-depth tour in Japan for Chinese tourists. I believe our quality of onshore sight-seeing as well as tourist experience will be greatly improved."

6.3 Important Enlightenment

As the gateway city of East China cruise market and the most important pivotal port of Northeast Asia, Shanghai has seen more and more international cruise lines settling down here. Plus the emerging local cruise brands, the competition in the East China market has been heated, along with such problems as product homogeneity and low-price competition. At the same time, with the constant development of the cruise market, tourists have increasingly deeper understanding of cruise tourism and multi-layer and further demands for their experience of cruise ships. Under this background, the development of Shanghai cruise market is entering a new stage and is in urgent need of truly differentiated and diversified cruise products.

How to develop in a healthy and ordered way and prevent homogenization and the low price problem has become a notable challenge in the fierce market competition. "Currently, our cruise market is growing rapidly, and certainly there are a lot of problems in this process. For example, there are only a few cruise routes and the facilities and activities are similar. Therefore, low-price competition exists in the market," said Zheng Weihang, Executive Vice President of CCYIA, during the interview. While such big and new ships as Norwegian Joy are entering the Chinese market, tourists will experience more activities during their journey, and they will see the cruise ships more like a destination than just a means of transportation. In the future, cruise lines and travel agencies may conduct precise positioning of tourists and differential operation, offering products of high end, middle end and low end and winning tourists with personalized service rather than lower prices.

Now a new pattern of home ports is taking shape in China, where East China is leading and the North and South China is supporting. Led by the home port of Shanghai, the East China market has been standing at the leading position for years. Despite the slightly slower growth since 2017, the Yangtze River Delta boasts unparalleled advantages in both economic level and popularization of cruise ships. Even if it is constrained by the homogenization of route products, major cruise lines can also keep tourists feeling fresh by designing theme routes and sending new ships. It can be foreseen that the East China market, represented by Shanghai, will still gather the most deluxe and novel cruise ships in Chinese market in the future. And as the less competitive cruise lines transfer to the North China or South China markets, the East China market will get out of the price war and maintain a certain ticket price level, which will be beneficial to the healthy development of cruise market. The North China market, supported by Beijing and Tianjin, although can hardly have good performance during winter due to climatic limitation, will still win much attention from cruise lines on summer routes and make contributions to the growth of Chinese cruise market. The South China market, thanks to its favorable climatic and geographical conditions, is adding vitality to Chinese cruise market with its prosperity. With its increasingly obvious market potential, there will be broader business areas and more diversified products. The new pattern will promote the healthy and sustainable development of Chinese cruise market, which has reflected that the Chinese market is gradually growing mature.

7 Hot Spot 7 Enhanced Industrial Policy and Research Support from National Ministries and Commissions Help Promote Stable and Further Development of Chinese Cruise Industry

7.1 Hot Event

On November 28, 2016, the General Office of the State Council released the *Opinions on Further Expanding the Consumption in the Fields of Tourism, Culture, Sports, Health, Old-age Care, Education, and Training*, which pointed out that a master plan for cruise tourism development should be set out. Meanwhile, customs formalities for cruise tourism should be normalized and simplified, and enterprises are encouraged to develop domestic and international cruise routes, so as to further promote the development of domestic cruise tourism. And the 15-day visa-free transit policy for foreign tourist groups that has been executed in Shanghai will be gradually extended to other cruise ports. On December 26, 2016, the State Council issued the *Tourism Development Planning for the 13th Five-year Plan Period*, which set a special column for "cruise and yacht tourism development planning" for the first time. It proposes the 5 requirements of "enhancing infrastructure construction, developing special routes, strengthening the subjects of cruise market, promoting cruise lease consumption, and cultivating cruise and yacht talents". Particularly, in terms of inbound cruise tourism, it proposes to promote the construction of visiting ports for international cruise ships, expand the applicable area of the 15-day visa-free transit policy for foreign tourist groups, expand the scope of pilot port cities in order, optimize inbound tourism policies, facilitate visa application and customs clearance for inbound tourism, and study and prepare facilitation policies for foreign cruise tourists. On January 12, 2017, the Ministry of Industry and Information Technology and the Ministry of Finance jointly published the *Action Plan of Deepening the Restructuring and Accelerating Transformation and Upgrading of Shipping Industry (2016–2020)*, proposing to accelerate the implementation of independent design and building of cruise ships. On May 19, 2017, the Ministry of Transport issued the *Action Plan of Further Promoting the Structural Reform of Water Transport Supply Side (2017–2020)*, putting forward 19 items including promoting the coordinated development of regional ports and accelerating the development of cruise and yacht transport. In July 2017, six departments including the Ministry of Transport and CNTA jointly issued *Several Opinions on Promoting the Integrative Development of Transportation and Tourism*, explicitly proposing to enhance the link between cruise ports and urban tourism system, guide qualified cities to build distributing hubs for cruise tourism, optimize the layout of coastal cruise ports and gradually build a reasonable cruise port system, encourage the addition of cruise routes, strengthen the construction of public service facilities for water tourism, etc.

In recent years, as China is promoting the rapid development of cruise industry, local governments also attach great importance to cruise tourism development, introducing a number of encouraging policies, laws and plans. From March to October 2016, Tianjin, Zhoushan, Qingdao, Sanya, Shanghai, Xiamen, etc. published local "Development Planning for the 13th Five-year Plan Period" successively, all of which encouraged the cultivation of local cruise industry and promotion of cruise industry development. In December 2016, Guangzhou City issued the *Several Measures of Accelerating the Development of International Cruise Industry in Guangzhou*, adding rewards for opening new cruise routes, expanding business of travel agencies, and starting new cruise lines. In January 2017, Hainan Province issued the *Implementation Measures of Financial Incentives for Encouraging the Development of Cruise Tourism Industry in Hainan (Trial)*, which classified cruise ships by tonnage and set out rewards for cruise lines, cruise ship operators and cruise ship leasers that open home port routes in Hainan (Haikou or Sanya). In February 2017, Qingdao Municipal Government published the *Opinions on Promoting the Healthy Development of Shipping Industry*, proposing to build Qingdao Port as a key international pivotal port serving "the Belt and Road", vigorously develop cruise economy, build cruise tourism destinations, build the "Cruise Center in North China" by 2020, and make Qingdao the "regional cruise home port in Northeast Asia". Meanwhile, as CNTA has approved to set up the "China Cruise Tourism Development Pilot Area" in Prince Bay, Shekou Industrial Zone, Shenzhen, the *Master Plan of Cruise Tourism Development of Shenzhen (2017–2030)* is now under preparation.

7.2 Comments on the Event

On the one hand, cruise tourism, as an emerging travel mode, has great potential and plays a more and more important role in driving economic growth and accelerating consumption transition and upgrading. On the other hand, as cruise home ports have strong driving effect for economy, many coastal port cities have been carrying out cruise port construction, which propels cruise industry development. It can be seen that, in order to better develop cruise tourism and improve cruise industry, national ministries and commissions as well as local governments have issued a series of supportive policies concerning tourism reform, cruise transport development, travel equipment manufacturing, etc. National level policies and regulations embody the top-level design, guiding the cruise industry development of the country; while local policies and regulations are operable and supplement the national policies, promoting the overall development in different places. Currently, these relevant policies have the following features: first, policies are developed according to the actual demand of cruise industry and cruise tourism development, e.g. the 15-day visa-free transit policy; second, there are more and more entities issuing policies related to cruise industry, not only limited to the State Council, but also local governments, NTA, transport commission, and even the State Development Bank, National Development and Reform Commission, etc.; third, the objectives are more and more diversified, not only targeted on economy, but also involving political and diplomatic objectives such as "the Belt and Road"; fourth, there are more diversified policy instruments, such as fiscal policy instrument, monetary policy instrument, environmental policy instrument, and exchange rate policy instrument. Of course, part of these policies still needs to be further refined. For example, the *Action Plan of Deepening the Restructuring and Accelerating Transformation and Upgrading of Shipping Industry (2016–2020)* proposes to organize and implement a batch of key innovative projects of large cruise ships, intelligent ships, and low-speed machine for ships, but this policy has few contents concerning large cruise ship manufacturing and is very general, so it can hardly guide the development of the industry.

7.3 Important Enlightenment

The extensive publishing of favorable policies for cruise industry at both national and local levels has dramatically propelled the rapid development and ensured the sustainable and healthy development of the industry. However, the following issues should be paid attention to during future policy making: first, supportive policies for inbound cruise tourism shall be actively developed. Currently, the domestic inbound tourism market is not performing well. It is high time that inbound tourism was promoted by international cruise ships to change this situation. To develop inbound tourism by cruise tourism is the development direction of cruise industry in order to make China a real tourism country. Active attempts should be made in tax-free shopping, etc. besides the visa-free policy. Second, relevant policies shall be extended to the whole industry chain. The development of each link of the cruise industry requires supporting policies. Now our national and local policies mainly focus on the midstream and downstream of the industry, but should be extended to the upstream, i.e. China should develop supplementary policies to encourage the development of cruise manufacturing. Third, relevant policies on external environment shall be developed. In the environment of market operation, cruise manufacturing, procurement, leasing and operation require huge amounts of funds and a developed financial service system for guarantee; meanwhile, infrastructure construction for the ports also needs large amounts of money. And with the rapid development of cruise industry, marine pollution problems are getting more prominent. The wastes discharged by cruise ships not only contaminate seawater and air, but also harm the health of coastal residents. Relevant authorities shall pay attention to the ecological environment of cruise tourism and introduce environmental protection policies to ensure the healthy and sustainable development of the industry.

8 Hot Spot 8 Academic Research of Chinese Cruise Industry Finds a Brand-New Perspective, Exploring the New Rules of International Cruise Industry in Depth

8.1 Hot Event

Our cruise industry has seen rapid development for over a decade. With the implementation of such fundamental policies as the "13th Five-year Plan", "the Belt and Road", and the structural reform of the supply side, the new situations of "New Normal" economic development and transformation and upgrading of manufacturing, and the continuous eastward-moving of the development center of global cruise industry, the cruise industry of China has seen bigger opportunities. Meanwhile, the academic researches about cruise industry have entered a more prosperous and in-depth stage.

Academic researches have three functions for cruise industry development: firstly, they analyze and solve the problems existing in industry development in a larger scope. The scholars are not stakeholders in the industry, so they can consider the overall interests of the industry from the standpoint of a third party, but not limited to the solution to individual problems. Secondly, the academic circle is the bridge between the industry and government departments. Industrial policies are the foundation and guarantee for cruise industry development, and the policies of both central government and local governments are made from the standpoint of the entire society. As for the demand for cruise industry policies, we need the academic circle to consider about the policy coordination from the perspective of the

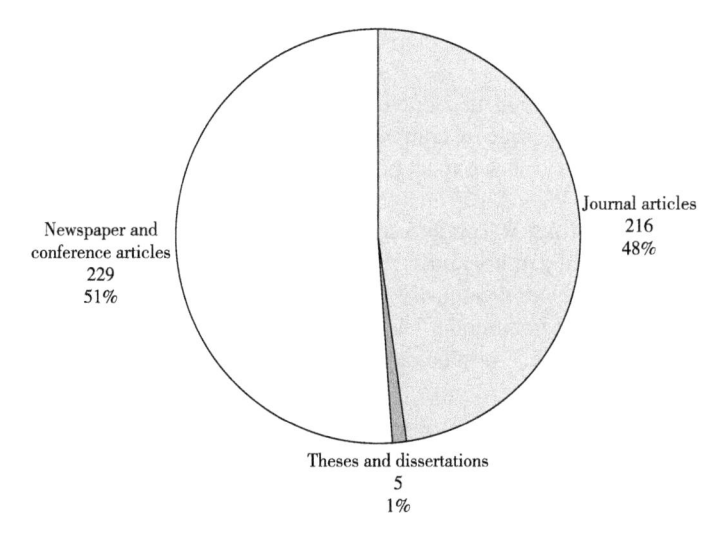

Fig. 1 Articles concerning cruise ships from October 2016 to July 2017

government. Thirdly, the sophisticated analyses on the development rules and mechanism of cruise industry made by the scholars from the theoretical level let the industry and enterprises better understand the development trend and better face the uncertainties in future.

From October 2016 to July 2017, there have been about 450 articles themed at "cruise ships" published on CNKI, ChinaInfo, and VIP journal resources, among which there are 216 journal articles, 5 theses and dissertations, 229 newspaper and conference articles. At the same period, there are 27 English articles containing the words of "cruise ship" or "cruise tourism" published on Web of Science (including SCI/CPCI) (see Fig. 1). The researches mainly focus on the five aspects of cruise ship building, cruise ship operation, cruise tourism, environmental impact and relevant researches:

8.1.1 Cruise Ship Building

Rajendran et al. (2016) studied the vibration caused by waves during the voyage and the vertical bending moment caused upon the entry into extreme sea areas by means of multiple experiments and mathematical analyses. Li and Lv (2017) obtained the characteristics of global cruise ship types by analyzing the comprehensive parameters of 228 cruise ships operating around the world in 2017.

8.1.2 Cruise Market

Sun and Ye (2016) verified that cruise products are involved in excessive competition featured by low price by comparing the ticket prices of domestic markets in 2014 and 2015, and held that it must be solved by differentiation and building new marketing channels. Wu (2016) analyzed the correlation and linkage mechanism between Shanghai Disney Resort and cruise industry. Cao and Ye (2017) combined the development trend and hot spots of cruise industry and proposed the directions worth studying for the cruise industry policy in future. Mei and Ye (2017) pointed out that our "chartered boat model" would be transferred to "cabin cutting model" and then to "retail model", continuously achieving all-win results and improving the experience of tourists.

8.1.3 Cruise Tourism

Sun and Hou (2017) summarized the sustainability and accountability of overseas cruise tourism in recent 15 years and proposed the theory and practice of "responsible cruise tourism". Niu and Shao (2016) and Cao and Ye (2017) combined the economic climate indexes and the development status of Chinese cruise industry to establish the economic climate indexes of Chinese cruise industry, which has provided the quantitative basis for researches on Chinese cruise economy. Liu et al. (2017) carried out a questionnaire survey on tourists in Xiamen and analyzed the consumption behavior characteristics of cruise tourism. Zhang et al. (2017) analyzed the path of sustainable development of cruise tourism with the online comments from ctrip.com. Sun et al. (2016) established the satisfaction evaluation indexes for home ports, surveyed the tourists departing from Shanghai, and proposed measures of improving the satisfaction toward the home port of Shanghai.

8.1.4 Environmental Impact

Li and Lv (2016) analyzed the features of cruise pollution, carried out quantitative estimation on the pollution caused by the operation in China, and proposed corresponding countermeasures for controlling the pollution. Ye et al. (2017) studied the impact of cruise ships on port environment, the building of overseas green cruise ports and domestic green cruise ships, and the path and measures for domestic building of green cruise ships. Ren et al. (2017) proposed to establish the cruise port environmental protection system of China based on relevant theories. Sun et al. (2017) selected the credit rating data of cruise ship environmental behaviors provided in the cruise ship environmental behavior reports published by FoEI (Friends of the Earth International) from 2011 to 2016 and analyzed the environmental behaviors of the cruise lines. They found that there were differences between such behaviors of different cruise lines and also between the behaviors of typical cruise lines in different areas. It is suggested that improvements be made in the aspects of cruise lines, government, cruise ports, social organizations and the public, to create a mutual-promoting and environment-optimization-oriented atmosphere for cruise industry.

8.1.5 Relevant Researches

Morillo et al. (2016) thought that Norovirus (No. V) was the most common cause of foodborne disease outbreaks on cruise ships.

Cui (2016) studied the job satisfaction of the crew members of cruise ships by questionnaire and comparative analysis. Chen et al. (2017) established the value assessment model for cruise tourists based on customer lifetime value theory. Tomaso et al. (2017) studied the risk control of cruise ships by means of cases. Ottoman et al. (2016) studied the burn care and disaster management on cruise ships, marine medical qualification, etc.

On the other hand, China has strengthened the support for the researches on cruise-related themes. Currently, there are five cruise-themed national scientific research projects, two of the National Natural Science Foundation of China and three of the National Social Science Foundation of China. The research team members come from traditional advantageous units, i.e. Shanghai University of Engineering Science, East China Normal University, Shanghai Maritime University, and Dalian Maritime University (see Table 1 for details). Besides, the first cruise-related international cooperation scientific research project of China, closely followed by China Communications and Transportation Association, i.e. the "Environmental Impact Assessment of Cruise Ship at Berth and the Ecological Benefits of Implementing Ashore Power Project to Reduce Pollution and Carbon Emissions from Ships" funded by the "Air and Water Conservation Foundation" of

Table 1 List of cruise-themed national scientific research projects in process

Name	Category	Description	Organization
Sun Xiaodong	2015 NSFC general project	Study on the influence factors, resource allocation and optimizing strategy of cruise route planning	ECNU
Xie Qimiao	2015 NSFC youth scientist fund	Study on the uncertainty analysis method for the evacuation time of large cruise ships in case of fire	SMU
Ye Xinliang	2016 NSSFC general project	Study on cruise tourism development of China under the structural adjustment of supply and demand	SUES
Yan Guodong	2017 NSSFC general project	Study on structure optimization of Chinese cruise industry in the perspective of all-for-one tourism	SUES
Guo Ping	2017 NSSFC general project	Study on the legal safeguard for cruise transport in the perspective of sustainable development	DMU

the National Geographic Society (the US), led by Dr. Sun Ruihong (Associate Professor of Shanghai University of Engineering Science), has been approved by the National Geographic Society and will come to a conclusion. This will actively promote the carbon emission reduction for the cruise ports of China and the US.

8.2 Comments on the Event

Domestic Researches: the academic circle mainly focuses on the midstream and downstream cruise operation and tourism, with the research objects of theoretical index creating and mechanism analyzing as well as problem evaluation and solutions. As for research method, empirical study and comparative study dominate. Since quantitative research requires large amounts of data and it is very hard to obtain the data from cruise operators, the academic researches of cruise industry are mainly qualitative, while quantitative researches mainly focus on tourists with accessible data, such as consumer behaviors and online comments.

Overseas Researches: English literature is much less than Chinese literature in number, but it involves a more extensive scope, including ship structure, food safety, risk control, and medical care.

Common Concerns: Xie Qimiao studied the evacuation on large cruise ships in case of fire, Ottoman et al. studied the burn care on cruise ships, and Tomaso et al. the risk management of cruise ships.

8.3 Enlightenments

First, compared with developed European and American countries, our cruise tourism has just started, and academic researches mainly focus on problem analysis during cruise operation. With the development of domestic and global cruise industries, academic researches will surely go deeper and wider, summarizing the development rules theoretically and providing more theoretical support for industrial development in a more active way.

Second, the cruise industry involves many aspects of many industries, such as national laws and regulations, technical innovation, talent cultivation, manufacturing, logistics, and service. It will exert a lasting and stable influence on our national economic development and play an irreplaceable role in continuous technical innovation and job creating.

Third, according to existing literature, the basis of both Chinese and English literature is mainly some theoretical frameworks and survey data, lacking cooperation with the industry. The summary of problem solutions and experience by scholars needs the participation of industry experts. The experts can provide basic data support for the scholars, while the research results of the scholars can provide support for solving the problems of the experts.

Fourth, the development of cruise industry is the foundation of academic researches, while the latter promotes the former. The industry focuses on economic benefit and tends to solve specific and individual problems. For example, if a cruise route of some cruise line suffers loss, the cruise line will analyze if there will any possible profit from this route; if not, it will be abandoned. The reason is that a certain enterprise or industry can only consider the solutions from aspects that are controllable by them.

9 Hot Spot 9 Exploring the Chinese Path of Cruise Distribution Model Under the Heated Debate on Cruise Distribution Model Innovation and Transformation

9.1 Hot Event

After a decade of rapid development, the cruise industry of China has entered the stage of in-depth market exploration and product upgrading. In the second decade of cruise industry development, how to innovate in cruise products and avoid homogenization and low-price competition, how to extend the channels to the inland and expand the passenger source, and how to establish new partnerships besides the chartered boat model have become the bottleneck and common concerns of the cruise industry. Under this background, channel providers and middlemen have strengthened communication, and there have been frequent forums and conferences about cruise channel distribution to discuss about the future channels of Chinese cruise industry. In 2016, there were heated debates about cruise market, price trend, product marketing and channel design during such conferences as Evergrand 2016 Travel Agency Cruise Distribution Forum, the 1st Private Advisory Board of Cruise Distributors of China, and the cruise distribution forum of Seatrade Cruise Asia Pacific. In 2017, the Princess Cruises 2017 National Cruise Distribution Conference was held in three cities of Beijing, Shanghai and Guangzhou, gathering major chartered boat operators and distributors, linking thousands of dealers, and building a communication platform between the cruise lines and the distribution channels of travel agencies.

9.2 Comments on the Event

As the Chinese cruise market has just emerged for a short time and is not mature enough, with incomplete marketing system and information management, cruise lines have limited contact to the Chinese market and will still rely on middlemen for market expanding, and the chartered boat model will continue to be the main marketing model. With further development of Chinese cruise industry, however, other marketing models will be tried as well. In 2017, Royal Caribbean

International has seen more and more non-chartered retail boat routes. In the past, cabin cutting totally relied on large customers, with the minimum cabin number of 200 or 300. However, now the minimum number is 8. Cruise lines like Princess Cruises and Costa still adopt chartered boat model as their main cooperation mode with travel agencies, but have reduced the proportion of wholly chartered boat and have been developing small and medium-sized customers, with flexible cooperation modes like chartered boat, half-chartered, large-scale cabin cutting, and retail.

Seen from the status quo of the Chinese market, several aspects need to be carefully considered about to establish a new cruise channel system: firstly, how can the authorization and price control range of large-scaled chartered boat, small-scale cabin cutting and retail provide equal and even greater privilege for chartered boat and cabin cutting while better motivating the retail agencies during the reform of chartered boat model? Secondly, in regional marketing, how can the cruise lines establish corresponding marketing systems in face of the regional development differences in the eastern and mid-western regions? Will the channel marketing targeted on new customers and the subsidies like transportation fee weaken the marketing effort on existing customer market? Thirdly, will the greater dependence on online marketing improve the situation of bundling-sale of current cruise products? Should the cruise ticket and contract price products be designed separately? Fourthly, should product publicity be enhanced in the market? How to balance road show activities that rely on the middlemen and product publicity targeted on the final consumers?

9.3 Important Enlightenment

First, to establish a broader and more efficient channel marketing system will be the main strategy of cruise lines for marketing in China. Chinese cruise market has huge space. With the continuously increasing supply, the market need to be arranged more rapidly, and the marketing and channels should be diversified.

Second, cruise lines need to achieve sincere cooperation with large chartered boat operators, online OTAs, etc. for win-win marketing. Cruise lines should realize further allocation and coordination with large agencies and underwriters, so as to achieve balance in the most efficient way and drive the market together. Provided that the inventory will be cleared and the cost be recovered rapidly by methods like chartered boat, the cruise lines can expand the marketing of single tickets by cooperating with Ctrip, Tongcheng (www.ly.com), Tuniu, self-owned Apps or sales terminals. In this way, the scale of chartered boat can be reasonably controlled, and the proportion of consignment sale can be increased to enlarge the profit space and carry out differentiated competition and adaptive positioning.

Third, cruise lines should establish a more flexible price system that is adaptive to the Chinese market. With the changes of cruise lines in the distribution system and layout in the Chinese market, how to handle the relations with chartered boat operators, OTAs, cabin cutting agencies, and small and medium-sized agencies will be more important for the cruise lines. And the focus of these relations is the wholesale price, cabin cutting price and agency price offered by the cruise lines. To establish a multi-layer price guiding system and adjust the price according to market changes will increase the anti-risk capability of the channels.

Fourth, while the cruise lines are adjusting their distribution structure, there must be corresponding support for channel marketing. If we hope to attract more and more small and medium-sized agencies to directly deal with cruise lines, we must have adequate policy support and corresponding marketing methods need to be changed, e.g. training for small and medium-sized agencies, various marketing activities involving terminal customers, necessary funds, and support from professionals. Emerging agencies are small, with poor anti-risk capability, especially in this kind of low-price competition environment. If all attention is paid to new customer developing but not to professional marketing training, these small travel agencies will easily disappear due to their loss.

Fifth, cruise lines should make customer files to realize overall marketing toward final consumers, and enhance follow-up service while ensuring their satisfaction onboard. Cruise lines should realize smart marketing via informatization, following existing tourists to maintain customer loyalty, and then achieve word-of-mouth marketing. In this way, they can build a second marketing channel without existing middlemen marketing system.

10 Hot Spot 10 The Nanhai Cruises Opens up Xisha Islands Route, Setting off on a Path to Off-Shore Cruise Ship in China

10.1 Hot Event

In December 2016, a new cruise ship, named Nanhai Zhi Meng, became a member of the family of Xisha tourism cruise route, sailing from Sanya to Crescent group of Xisha Islands. Destinations include Quanfu Island, Yagong Island and Yinyu Island. "Nanhai Zhi Meng" was built by Guangzhou Shipyard International Company Limited with the total tonnage of 24,572 t and the maximum capacity of 893 people. It is now under the operation and management of Sansha Nanhai Dream Trip Cruises Co. Ltd. In March 2017, a new cruise ship "Princess Changle" for ecological tourism route on Xisha Islands set sail for the first time at Sanya Phoenix Island International Cruise Terminal. Tourists may take a two-day trip on Xisha Islands, enjoying views of 3 islands: Quanfu Island, Yagong Island and Yinyu Island, and joining theme activities such as raising the national flag, environmental protection volunteer, semi-submersible ship sightseeing, snorkeling and sea fishing.

10.2 Comments on the Event

The Nanhai Sea covers an area of 3.56 million km^2, and it is the deepest and largest sea in China with the average water depth of 1212 m and annual average temperature of 25–28 °C. Developing Nanhai Cruises tourism exerts profound influence on building China into a maritime power, building the "21st Century Maritime Silk Road" and pushing forward the development of Hainan international tourism island, driving our marine tourism into deep blue development. Xisha is one of the purest virgin lands, a perfectly pure sea area at the edge of the Pacific Ocean and the most beautiful sea area in China. With strong mysterious color and incredible charm, it is the southernmost destination for Chinese tourists to approach currently and also known as the most difficult place in China to access.

Tourist will experience the best sky, sea, islands and beaches. It is the paradise that tourists desire. The strong attraction of the South China Sea to local cruise lines lies in the strong interest from tourists, protective policies, great advantage in maritime resources and the vast and wonderful underwater world. Due to the vulnerability of its original ecology, small and medium-sized cruise ships are particularly suitable for the development of cruise tourism in Nanhai Sea area. Island development, cruise tourism without a destination, sea fishing and diving are all great activities, but Nanhai Sea area is not fit for a great number of large ships to enter into in a large scale. It is the high-end market area to be developed by Chinese cruise enterprises. Nanhai cruise tourism is of a relatively small scale, leaving open the possibilities of restrictions regarding policies, extreme weather and infrastructures. However, the Nanhai Sea region will still become the best destination to develop cruise tourism with strong attraction to tourists, which makes it the golden place of Chinese cruise tourism development and where the unique advantage of diversified competition among Chinese cruise enterprises lies in.

10.3 Important Enlightenment

China has a long coastline extending from the Yalujiang Estuary in the north to the Beilun Estuary in the south, a total of more than 18,000 km. Coastlines of islands reach a total of over 14,000 km, including more than 400 km natural deep-water lines. China's waters include the Bohai Sea, the Yellow Sea, the East China Sea, the South China Sea and waters east of Taiwan, of which the Bohai Sea has a mainland coastline of over 2700 km, and those of the Yellow Sea, the East China Sea and the South China Sea are more than 4000, 5700 and 5800 km respectively. A plenty of coastal resources provide favorable basic conditions for off-shore cruise tourism. Additionally, coastal cities such as Dalian, Qingdao, Shanghai, Xiamen and Sanya are famous tourism destinations with abundant tourism resources and cruise port resources and are major tourist origins in cruise market, providing favorable conditions for off-shore cruise tourism.

Nanhai cruise tourism will strongly push forward the development of Chinese local cruise tourism and improve the diversified competitive advantage of Chinese cruise enterprises. To develop Chinese local cruise tourism, it requires Chinese cruise fleet and the planning and design for Chinese cruise routes. The implementation of differentiation strategy is necessary for the development of Chinese local cruise tourism. The differentiation should be embodied not only in the hardware design for cruise ships as well as products and services characterized of Chinese culture, but also in the planning and design for routes.

As an up-rising star, Chinese local cruise enterprises are difficult to be absolutely dominated. In contrary, small and medium-sized cruise ships are more suitable for the development of Chinese cruise fleet. Cruise ships with 20,000–100,000 tonnage are the major force of Chinese cruise fleet, among which those with 20,000–50,000 tonnage target high-class customers, and those with 50,000–100,000 tonnage target medium and high class customers—mainly young white collars. Focus should be placed on enterprise customers to attract a lot of commercial activities as well as meetings and ceremonies to be held on the cruise ship. Enterprise customers produce more economic efforts than individual tourists do, and attract more advertisers at a lower market development cost.

Chapter 3
Active Connection to "The Belt and Road" Initiative to Enable the Great-Leap-Forward Development of Shanghai Cruise Economy

Hong Wang, Xinliang Ye and Shuang Cao

In September 2013, China first proposed to jointly build the "Silk Road Economic Belt" in Kazakhstan, in October the same year, indonesia proposed to step up maritime cooperation with all ASEAN countries to jointly build the 21st century "Maritime Silk Road", by now, the "Belt and Road Initiative" was officially proposed. In March 2015, the China Development and Reform Commission, the Ministry of Foreign Affairs and the Ministry of Commerce jointly issued the document `Promoting the Vision and Action of Building the Silk Road Economic Belt and the 21st Century Maritime Silk Road', which further specified the direction for the development of the Belt and the Road initiatives.The National Development and Reform Commission, Ministry of Foreign Affairs of the People's Republic of China and Ministry of Commerce of the People's Republic of China jointly issued the document titled *Vision and Actions on Jointly Building Silk Road Economic Belt and 21st Century Maritime Silk Road*, further indicating the development direction of "the Belt and Road" initiative.

Hong Wang—Research field: Social assurance, Cruise tourism and marketing strategy.
Xinliang Ye—Research field: Cruise tourism and risk management.
Shuang CAO—Research field: Cruise tourism.

H. Wang (✉)
Shanghai International Cruise Business Institute, Shanghai, China
e-mail: wh@sues.edu.cn

X. Ye · S. Cao
Shanghai University of Engineering Science, Songjiang, China

© Social Sciences Academic Press and Springer Nature Singapore Pte Ltd. 2018
H. Wang (ed.), *Report on China's Cruise Industry*,
https://doi.org/10.1007/978-981-10-8165-1_3

1 Background and Significance of Research

1.1 Accelerated Integration of "The Belt and Road" Initiative with Cruise Industry

The first 46-day South Pacific Rim cruise route departing from the home port in China cunningly developed by Caissa Tourism has become an important practice of the Maritime Silk Road into tourism. In June of the same year, the ministerial meeting for the Silk Road tourism approved *Tourism Ministerial Meeting of Countries along the Silk Road Economic Belt*. In July 2016, on the Maritime Silk Road Financial Summit Forum 2016 and Hainan Finance Expo, it is proposed to open cruise routes and non-stop air routes between Hainan and counties/regions along the Belt and Road, and to promote the multi-stop-on-one-trip cruise tourism with joint efforts of national and international major cruise port cities. With the help of island touring policy forum, endeavor should be made to form the tourism union for the 21st Century Maritime Silk Road. In September, the first session of Silk Road (Dunhuang) International Cultural Expo was organized with the theme of promoting cultural exchange and seeking cooperation and development. In November, the working session 2017 of Chinese Silk Road Tourism Promotion Union was held in Haikou City of Hainan Province to discuss the development plan for the Maritime Silk Road. In January 2017, the Silk Road Tourism Belt was included in the 13th Five-Plan for tourism. In May 2017, the Belt and Road Forum for International Cooperation was held in Beijing on which China entered into several agreements with international parties on tourism cooperation. In June, Xi'an Silk Road Culture Tourism Research Association signed a Memorandum of Understanding with Kazakhstan Tourism Association on cultural tourism research.

1.2 Active Participation of Port Cities in the Construction of 21st Century Maritime Silk Road

The proposal of the Belt and Road development strategy brings a lot of development opportunities for tourism—cruise industry in particular. Port cities also seize the opportunity to promote port development and accelerate the integration into the construction of the 21st Century Maritime Silk Road.

As the first home port city in China, Shanghai takes an active stand on responding to national development strategy by establishing the first Experimental Area for China Cruise Tourism Development in China. It defines the working target of accelerating the formation of cruise industry chain and formulates the action plan for cruise economic development integrating the zone and ports, so as to fully push forward building Shanghai into an international cruise city. Additionally, it cultivates the cruise industry chain, innovates new patterns of Baoshan international cruise services, and expands spaces for cruise development, striving to build the cruise industry into a pioneering industry for the Maritime Silk Road.

Fujian Province signs cooperation agreements with ports along the Belt and Road to strongly impel its ports to play a greater role in constructing the Belt and Road and the Maritime Silk Road; to channel more investment in port infrastructure construction and effectively improve the carrying capacity; to deepen the participation of countries along the Belt and Road in port development and construction, unceasingly expanding back-lands at coastal ports; and to compete for the implementation of "Early and Pilot Implementation" policy. The State Council approves to launch a pilot program in Pingtan County regarding sea transportation to Taiwan. The Ministry of Transport of People's Republic of China authorizes the provincial Department of Transport to issue the *Administrative Measures for Road Transport in Fujiang Province and Taiwan*, and the *Tentative Administrative Measures for Road Freight Transport between Fujian Pingtan County and Taiwan*, enabling increasingly close connection between Fujian and Taiwan. Fujian and other countries/regions in Asia, Africa as well as Hong Kong, Macao and Taiwan have complementary advantages in terms of resource composition, industrial structure and commodity trade, indicating a wide vision of cooperation in port shipping services.

As a port of departure, Guangzhou also takes advantage of the opportunity of national construction of the Belt and Road to develop hub port economy. It opens the sea routes with the destinations of Hong Kong, Vietnam and Japan. In *Several Measures for Accelerating the Development of Guangzhou International Cruise Industry*, a plan is proposed for building Guangzhou into one of largest cruise home port in Asia by 2020, and building Spratly into an international navigation center.

Shandong Province issues the *Opinions on Putting the Document No. [2014] 32 about Promoting Health Development of Shipping Industry Issued by the Development Research Center of the State Council into Practice* which specifies it will perfect cruise services at ports in Weihai City, Qingdao City, Yantai City and Rizhao City, gradually forming a reasonably distributed and well arranged cruise port system integrating home port, port of departure and visiting port. It also delineates the route map for Shandong to develop shipping industry, which is helpful to create an internationally competitive environment for shipping industry development, providing important opportunity for related enterprises to integrate into the construction of demonstration area for the Belt and Road strategy and local economic cooperation in Sino-Korea Free Trade Zone.

Guangxi Province supports a series of cities such as Fangchenggang to accelerate the development of cruise tourism; to promote the construction of cruise service infrastructures; and to accelerate the upgrading of berth and joint port inspection hall, trying to open international cruise routes from Fangchenggang to Ha Long Bay and to Da Nang in Vietnam as scheduled, thus achieving the great leap in the construction of the Maritime Silk Road and pushing ahead with the development of "the Belt and Road" cruise tourism.

Haikou City strengthens its industrial cooperation with Philippine. The local government delegation negotiate successively with Philippine Ports Authority and

CITIC Cruise Line regarding opening the cruise tourism from Haikou to Manila, which facilitates the development of cruise industry in Haikou and give full play to its advantage as the pivot city on "the Belt and Road".

2 Literature Review

With "the Belt and Road" being the center of attention, academic circles also start preliminary research on the connection of "the Belt and Road" with tourism and with cruise industry. Search the keywords "the Belt and Road", Maritime Silk Road and Cruise Tourism on www.cnki.net, you will see the network chart as shown in Fig. 1. Evidently, scholars focus their researches on industrial development strategy, marine economy and port cooperation.

After the presentation of "the Belt and Road" initiative, Luo [1] was the first to carry out the research on impacts of the Belt and Road strategy on cruise

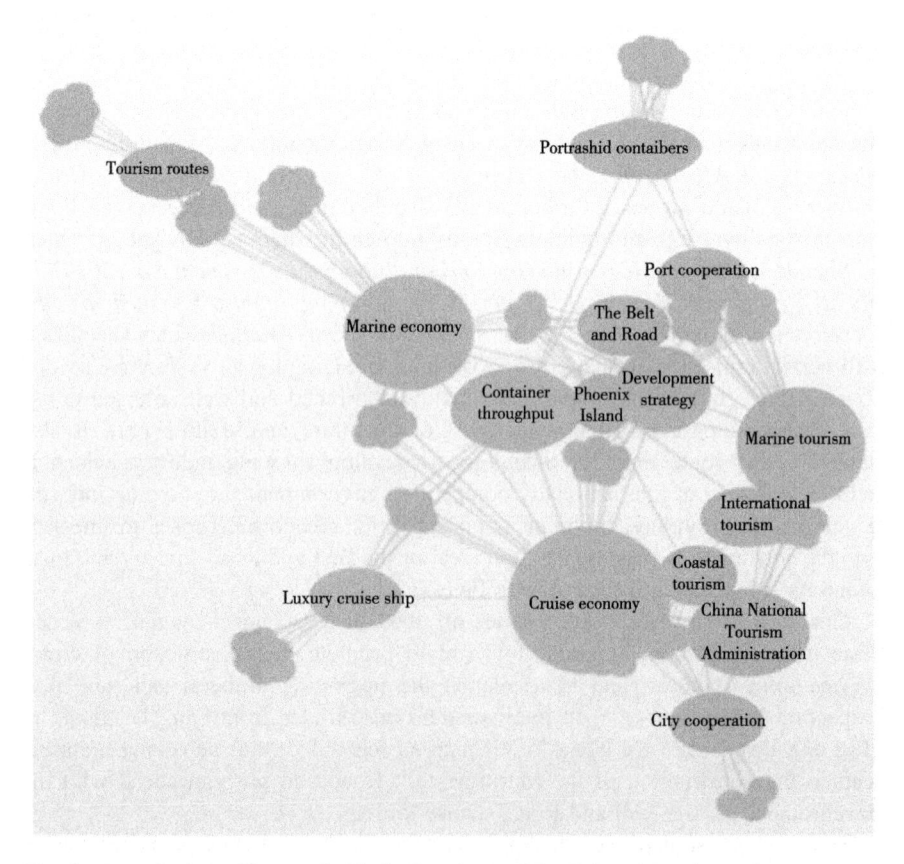

Fig. 1 Network chart of keywords (Cruise Tourism and the Belt and Road) coexistence

industrial development, analyzing the status and opportunities of cruise industrial development in Fujian case study and recommending Fujian to accelerate the integration into the 21st Century Maritime Silk Road construction. Soon afterwards, Zhang (2017) analyzed the cruise tourism market in Sanya City by means of PEST model to identify the issues existed in Sanya cruise tourism market and made a series of recommendations including strengthening the promotion of cruise tourism, uncovering cruise tourism safety secrets, cultivating cruise professionals, developing new routes relying on the Maritime Silk Road and opening more onshore sightseeing routes.

In the context of the Belt and Road strategy, the largest development opportunity and breakthrough for cruise industry is to establish long-term partnership with countries/cities along the Belt and Road and to develop industrial cluster accelerating the industrial development. Lu et al. (2017) determined the degree of maturity of supporting facilities for cruise industry in Xiamen City and its surrounding regions by means of location quotient, analyzed the potential of developing cruise industrial cluster in Fujian Province, and recommended to develop cruise-related industries with Xiamen being the center, Xiamen-Zhangzhou-Quanzhou being the sub-center, promoting the development of cruise home port in Fujian. Deng and Chunxiang [2] analyzed the status and issues of the tourism cooperation between Hainan Province and ASEAN, and pointed out five advantages for this cooperation —location, platform, human geography, industry and routes. Finally, recommendations were made in six aspects—tourism cooperation mechanism, utilization of overseas Hainanese resources, cruise cooperation enhancement, differentiated partnership implementation, exertion of comparative advantages and tourism education enhancement.

Cruise economy and the Belt and Road strategy are hot topics and develop rapidly, driving the development of many industries. How is the relationship between them? Ma et al. [3] studied such a relationship in Xiamen case study.

Starting with cruise industry development status, he made five recommendations for cooperative advantages under the political and economic background for cruise industry to be integrated into the Belt and Road construction, including the improvement of infrastructures, exploration of new routes relying on geographical advantage, enhancement of partnership with travel agencies, strengthening of partnership with economic circle and the enhancement of talent education assurance system and talent fostering.

With the rapid development of cruise industry, talent demands become increasingly higher. The presentation of the Belt and Road initiative enables the improved partnership with countries along the Belt and Road while it is pushing forward the flow of hi-tech talents, thus facilitating the exchange of talent fostering methods. However, the studies on cruise talents are less. The only literature available was from Yao (2016), which analyzed employability and marine tourism features, recommending that cruise talents should have marine skills, relevant job operational experience, language skills and thoughts.

Scholars' researches on the combination of Belt and Road strategy with cruise industry are mainly from the view of marine economy and cruise economy as its results, and secondly from partnership of port cities along the Belt and Road. So single is the research view and so limited are the contents that those researches are far from meeting demands from national cruise industry for rapid development.

3 Object and Goal of Research

On basis of the current hot spot of study, taking Shanghai cruise industry as the major object of study, this paper analyzes, by referring to a great number of academic literature and news reports as well as interviews of senior experts and insiders and field visits of Shanghai Wusongkou International Cruise Terminal, new opportunities that the Belt and Road development strategy brings to Chinese cruise industrial development. Furthermore, by taking into account the political and economic environment for the development of Shanghai cruise industry, it discusses how Shanghai cruise tourism connects with the Belt and Road initiative so as to enable the great-leap-forward development of Shanghai cruise economy.

4 The Development Opportunity for Chinese Cruise Industry Under the Belt and Road Strategy

At present, the rapid development of China's tourism industry has enabled the tourism industry to fully integrate into the national strategic system and become a strategic pillar industry of the national economy. The scale of outbound tourist volume and the tourism consumption are the highest in the world, and the cooperation with various regions and international tourism organizations has been continuously strengthened. Impelled by the great development and prosperity of Chinese tourism industry, China has become the first cruise market in Asia and Pacific area, and the second in the world. As a highly global, dynamic and open emerging industry, the distinctive advantage is that cruise tourism takes the lead to make connection. The Belt and Road strategy brings a lot of opportunities for the development of Chinese cruise industry.

4.1 The Belt and Road Provides New Development Concepts for Cruise Industry

Cruise tourism is characterized of innovation, coordination, openness and interaction, environmentally friendliness as well as joint development and shareability. Five new development concepts, namely innovation, coordination, greenness, openness and shareability, facilitate the improvement of development concept for international cruise industry. The development of cruise industry transforms from relying upon element-driven pattern to innovation-driven pattern, which means it changes from relying upon capital, hardware construction and lands to be dependent on technical, managerial and institutional innovation. Following the Silk Road spirit of being peaceful and cooperative, open and inclusive, happy to learn from each other and mutually benefit, partnership is established with nationally and internationally reputable cruise tourism cities for mutual development, in the aim of creating a community of international cruise development, thus jointly pushing forward the innovative development of global cruise industry.

4.2 The Belt and Road Provides New Development Patterns for Cruise Tourism

At present, China is during the transformation from cruise tourism to cruise industry and from low-end industry chain to high-end industry chain. It will complete the transformation from traditional industry chain to innovative chain and value chain, thus continuously improving its creativity. The Belt and Road development brings about the transformation of Chinese cruise industry from "fighting alone" to "developing together", and the effective connection with governments, associations, cruise-related enterprises and cruise research institutions in countries along the Belt and Road, so as to facilitate the collaborative development of the Belt and Road international cruise organization, enable the effective connection with cruise industry development strategies and policies, and strengthen the integration of resources and funds. To create a favorable environment for industrial development with joint efforts is helpful to foster new development patterns and formation of cruise industry.

4.3 The Belt and Road Provides New Development Motivation for Cruise Industry

At present, the world is seeing very rapid development of cruise tourism with the compound annual growth rate of 8% which is continuously higher than the global economic growth rate and traditional tourism growth rate. The globalization of cruise market and internationalization of tourism competition become intensified. The Belt and Road, particularly the 21st Century Maritime Silk Road provides it with innovative development motivation. Chinese cruise tourism has raised its international position increasingly. The tourism amount at home ports ranked the second in the world in 2016. Cruise tourism plays an important role in promoting concept propaganda and internationally cultural spread, cooperating with national diplomacy strategy and promoting peaceful and mutually benefit development in the world, revealing its bridging effect.

4.4 The Belt and Road Provides New Development Space for Cruise Industry

Cruise tourism development has strong cross-regional feature, making it a new window for promoting friendly partnership among countries along the Belt and Road and a new path to pushing forward the international cooperation and joint development of tourism industries in countries along the Belt and Road. China's cruise tourism is currently in a period of route adjustment and diversification.

Cruise ships are allowed to navigate from Chinese coastal ports to the Indian Ocean (or even Europe) and the South Pacific Ocean through the South China Sea.

5 Development Status of Shanghai Cruise Industry

5.1 Shanghai Cruise Industry Economy Grows Continuously to Raise Its International Position

As an intersection of the 21st Century Maritime Silk Road and Yangtze River Economic Zone, Baoshan District has become the most important water gateway in Shanghai and the international passenger transport port as well as the location of the largest cruise home port in Asia. In 2016, 471 cruise ships/times were received as well as 2.84 million inbound and outbound tourists. In the first quarter of 2017, cruise terminals received 104 cruise ships/times, representing a year-on-year growth of 33%, including 99 ships loaded or unloaded at home ports, representing a year-on-year growth of 38%. The throughput was 593,824 passengers, representing a year-on-year growth of 27%. Now, Wusongkou International Cruise Terminal in Baoshan has become the cruise home port in Asia that receives the most tourists (Figs. 2 and 3).

5.2 Baoshan Puts Great Efforts into the Development of Cruise Economy to Be the first to Create Cruise Industry Chain

With the upstream position, Wusongkou International Cruise Terminal, together with China State Shipbuilding Corporation (CSSC), China Investment Corporation, Fincantieri S.p.A., Carnival Corporation & PLC and Lloyd's Register of Shipping,

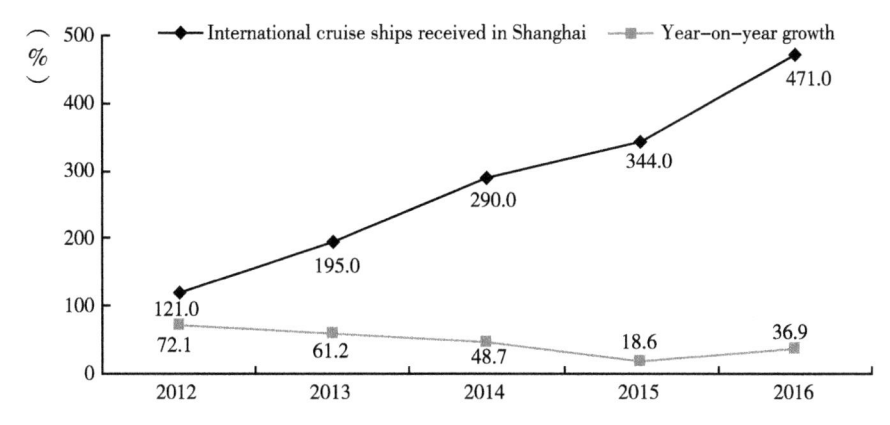

Fig. 2 Trend of international cruise ships/times received in Shanghai between 2012 and 2016

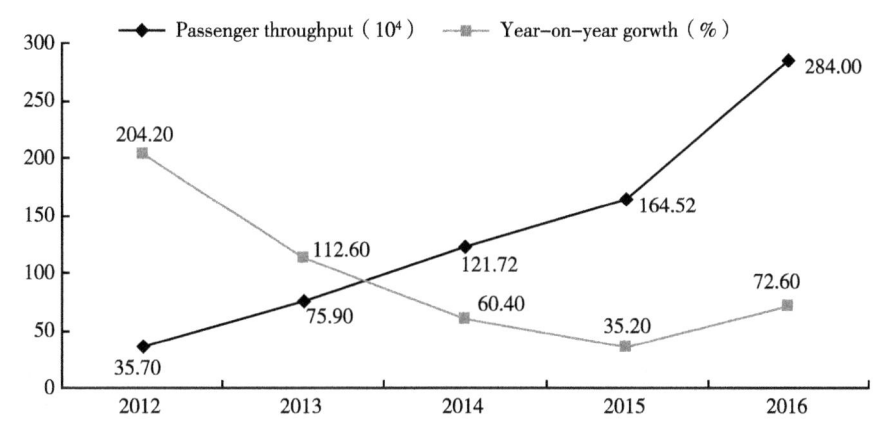

Fig. 3 Trend of inbound and outbound passengers received in Shanghai between 2012 and 2016

forms a union for native luxury cruise industry development, trying to build a system for supporting industries for cruise construction. At the midstream, by continuously improving investment attraction service level and deepening the partnership with large cruise enterprises, success is made in importing more than 30 cruise enterprises including Costa Cruises and MSC Cruise. At the downstream, by making innovation in launching services such as cruise insurance, cruise "through train" and convenient barcode for cruise clearance, establishing "Wusongkou" cruise services brand, which boosts Baoshan economy. Baoshan gradually steps from the City of Iron into the City of Cruise, continuing its glorious history of Gateway on Oceans.

The first round of cruise industry fund (30 billion yuan) is registered in Shanghai CSSC International Cruise Industrial Park. In the future, this fund will focus on cruise design and building, investment operation and supporting services for project funding, which helps to promote and foster Chinese native cruise building industry chain and attract more cruise supporting services enterprises to register at the park, providing impetus to Shanghai cruise industry growth acceleration. In addition, the subsequent works of Wusongkou International Cruise Terminal costing a total of 800 million yuan commenced in 2015. After the expansion, the total shoreline length is expected to extend from 774 to 1600 m, with 4 large cruise berths. The entire project is expected to start commissioning this year and to develop the capacity for accommodating 4 ships by 2018. The aims during the 13th Five-Year Plan period include that 20–30 cruise ships end their journey at this terminal, 800–1000 ships/times departed from this terminal carrying 5–6 million passengers; the amounts of cruise ship berth and tourist reception rank among the top three in the world; over 20 additional cruise institutions are established, thus continuously accelerating Baoshan's transformation from being a traditional industrial district to being a modernized and internationalized city.

5.3 Wusongkou International Cruise Terminal Leads Chinese Cruise Industry into an Era of Big Ships While It Is Making Leap-Forward Development

From the determination of site selection to formal operation, it succeeds in increasing the number of voyage from 60 to 471 and the tonnage from 70,000 to 160,000. The year 2017 greets Norwegian Joy, Superstar Virgo, Majestic Princess of Princess Cruises, the first ship customized for Chinese market. Many large cruise ships with Chinese elements, one after another, choose Wusongkou International Cruise Terminal as home port, driving Chinese cruise market to soar.

6 Recommendation on the Connection of Shanghai Cruise Industry with the Belt and Road Strategy

Cruise industry plays an active role in facilitating organized and free flow of navigation, efficient resource allocation and deep integration of cruise market in countries along the Belt and Road, promoting these countries to implement open-up policies and effecting regional cooperation to a greater extent, higher standard and deeper level. To push forward the effective connection of Shanghai cruise industry with the Belt and Road strategy is a necessary path to the realization of stable and further development of Shanghai cruise industry, the enhancement of partnership with international cruise industries and the improvement of international competitiveness; it is an important path to accelerating the development of Shanghai cruise industry chain, innovation chain and value chain, and fully improving the innovative development of cruise industry. To better connect with the Belt and Road strategy, recommendations are made as follows for Shanghai cruise industry.

6.1 Promote the Internationalized Application of Port Advantages and Standards Along the Belt and Road

Cruise ports play an important and fundamental role in cruise industry development. Over the decade, Shanghai cruise ports have improved hardware facilities, accumulating abundant infrastructure construction experience and operational management experience as well as strong technical strength. The construction of international cruise home port will become a key aspect of "the Belt and Road"—base tourism. Countries and regions along "the Belt and Road" and the 21st Century Maritime Silk Road own abundant maritime resources which are ideal for developing the cruise industry. Cooperation will be continued with AIIB, New Development Bank, the World Bank and other multi-lateral development organizations to support Chinese enterprises to construct cruise terminal infrastructures in countries along "the Belt and Road" and promote regional advantages of China and the internationalized application of terminal standards. Shanghai is the leader of the cruise industry that can accelerate the output of intelligence and talents as well as operation and management experiences in China's cruise terminals to countries and regions along "the Belt and Road". Efforts shall be made to complete the educational and training system for the cruise industry and to form Chinese standards in the international cruise industry.

6.2 Promote the Development of Cruise Products Along the Belt and Road

56 Countries and regions along "the Belt and Road" will become the destination of outbound tours of Chinese citizens. Promote the "multi-port call" policy along "the Belt and Road" and diversify cruise routes of the world. Cruise tours are usually international and trans-regional and thus can integrate tourist resources, destinations and ports along "the Belt and Road". The promotion of the "multi-port call" policy is an important path of development and can diversify the world's cruise routes. Open the boutique cruise routes along the Belt and Road with distinctive regional characteristics, so as to accelerate the exploration of long intercontinental and global voyages with Shanghai being the home port and visiting port.

6.3 Promote the International Marketing Promotion Along the Belt and Road

The tourism industry can enhance cultural exchange and civilization integration of people along "the Belt and Road". China has become an important destination for outbound tours of residents of many countries along "the Belt and Road". According to the statistics of CNTA, it is expected that during the 13th Five-Year Plan, China will attract 85 million international tourists from countries along the Belt and Road, stimulating tourism consumption amounting to USD110 billion. According to the report on inbound tourism provided by CTRP, about 14.5% of inbound tourists come to Shanghai per year, and it is predicted on this basis that Shanghai tourism consumption will be as high as USD15.95 billion. Cruise tourism, as a high-end travel mode, has developed rapidly in China with China as the home port, but the inbound tourism has been in slow pace. As an important water gateway and the location of international passenger transport port and the largest cruise home port in Asia, Shanghai needs to actively seek partnership with countries/regions along the Belt and Road, push forward the construction of such a tourism brand as Cruise Tourism—Maritime Silk Road, jointly hold large promotion events such as 21st Century Maritime Silk Road Tourist Year and Silk Road International Cruise Tourism Festival, thus raising the reputation and influence of Chinese cruise tourism along the Belt and Road and Shanghai Wusongkou International Cruise Terminal.

6.4 Promote the Interconnection of Cruise Industry Development Along the Belt and Road

"The Belt and Road" initiative has provided a new development platform for the international cruise industry that effectively interconnect China with the western cruise market. The high openness of the industry needs the support from the think tank as well as international exchanges of the industry, academic circle, government agencies and industrial associations. As an intersection of the 21st Century Maritime Silk Road and Yangtze River Economic Zone, Shanghai should be active to develop relative unified rules and standards for international cruise industry development together with countries along the Belt and Road, establish a high-quality exchange platform for international cruise tourism with joint efforts, and establish strategic partnership with cruise port cities. It should also strengthen communication in regard to cruise clearance, supervision, information, data and procedures on collaborative basis. By simplifying process cost of cruise development, furthering mutual recognition of law enforcement, data, information and standard sharing, active efforts should be put into the exploration of new patterns and approaches of interconnected supervision; it should promote convenient visa procedure, particularly attaining visa-free entry for Chinese cruise tourists to more countries. Promote traffic right opening, mutual recognition of certificates and tourism insurance.

6.5 Carry Out Researches on Talents Fostering Programs for International Cruise Tourism Along the Belt and Road and the Related Subjects

Professional talents are the primary resource to develop and expand the cruise industry. The cruise industry is a highly combined and international industry that requires highly internationalized talents who have global perspectives and views. The need for cultivating and attracting such talents is urgent. To this end, Shanghai should provide support for talent cultivation for cruise industry development in countries along the Belt and Road. Under the unified planning by Ministry of Education and Shanghai Municipal Government, universities/colleges with plentiful cruise expertise, for example, Shanghai University of Engineering Science may attract students from countries along the Belt and Road to study cruise expertise for bachelor's and master's degree.

Meanwhile, Shanghai should actively push forward the formation of multinational and trans-regional teams for international cruise industry study along the Belt and Road. During the development of cruise industry in countries and regions along the Belt and Road, certainly there will be many questions for study. It is necessary to give full play to the expertise and influence of Shanghai International Cruise Business Institute, hold conferences on cruise industry along the Belt and Road, establish the association of international cruise industry along the Belt and Road, and formulate new standards for international cruise industry along the Belt and Road.

7 Conclusion

The presentation of the Belt and Road development strategy brings new development concept, new development motivation, new development space and new development pattern for cruise industry. As the cruise port city with best development advantage and most development potential, Shanghai has high-quality development resources, superior geographical location and good development environment, and therefore should be more active to connect with 21st Century Maritime Silk Road construction in the era of opportunity. It should seize the opportunity to push forward the internationalization of native cruise industry with its advantages; make innovation of cruise routes with its port resources; push forward the international marketing of native cruise tourism with cultural exchange platform; actively reach friendly partnership agreements with countries along the Belt and Road so as to facilitate the interconnection of cruise industry; focus on studies on international flow of talent cultivation and related subjects, so as to achieve the leap-forward development of Shanghai cruise economy and promote all-round upgrading of Shanghai cruise industry.

References

1. Luo. Q. (2014). Fujian to speed up the port development, "Hays" strategic hub [J]. *Fujian Textile, 2014*(08), 24.
2. Deng, Z. P., Chunxiang, L. (2017). Research on Tourism Cooperation between China and ASEAN under the Belt and Road Initiative [J]. *Value Engineering, 36*(27), 237–239.
3. Malong, J., Hong, X., Shaotiong, C., & Zheyu, Z. (2015). "The Belt and Road" perspective on the development of Xiamen's cruise ship economy [J]. *China Water Transport (second half), 15*(10), 103–104.

Chapter 4
The Development and Changes of Cruise Industry in China Over the Decade

Jianyong Shi, Xinliang Ye and Junqing Mei

1 Introduction

Cruise tourism firstly originated in Europe and America. With the cruise tourism markets in Europe, America and other developed regions becoming increasingly saturated after decades' development, cruise lines begin to shift their development focus to the Asia-Pacific region which has 1/3 population of the world and is experiencing vigorous economic growth, instilling new energy to the development of tourism, promoting further development of maritime tourism and providing new momentum for economic development of the region. With the rapid economic growth, increased purchasing power of residents and great demands for tourism, China has become a key target market for international cruise lines which are already looking to tap the potential Chinese market by investing new luxury cruise ships. With a decade's development, China's cruise market has been expanded and developed more market potential, with the trend of development going far beyond expectation on the basis of rapid growth from over 10,000 tourists in 2006 to over 2,000,000 tourists in 2016. Cruise tourism has shown strong attraction and vitality in China. At present, China has become an important tourism destination in Asia-Pacific cruise routes. Being an integral part of the global cruise industry, the cruise industry of China is showing its greater international influences, and playing an important role in the global market.

Jianyong Shi—Research directions: Cruise tourism, Oriental management, Marketing; MEI Junqing, Graduate student of Shanghai University of Engineering Science.

J. Shi (✉) · X. Ye · J. Mei
Shanghai International Cruise Business Institute,
Shanghai University of Engineering Science, Shanghai, China
e-mail: shijy@sues.edu.cn

© Social Sciences Academic Press and Springer Nature Singapore Pte Ltd. 2018
H. Wang (ed.), *Report on China's Cruise Industry*,
https://doi.org/10.1007/978-981-10-8165-1_4

2 Literature Review

2.1 Overseas Research

Cruise tourism developed early in the global market. Europe and America have developed orderly development patterns and conducted a series of research on cruise industry.

Regarding economic contribution of cruise tourism, Mescon and Vozikis (1985), Braun et al. (2002) evaluates the contribution of cruise economy in Miami, the Cruise Capital of the World, and Canaveral by using an input-output model. The results show that cruise tourism has positive effects on the development of cruise ports and local economy, and the results show how the couise tourism contrioutes to the local economic growth. Diakomihalis tries to provide an in-depth analysis of how the cruise tourism stimulates the economic development of Greece by adopting the Tourism Satellite Account. However, difficulties encountered during the actual data collection due to the complexity and the diversity of data resulted in troubles in implementing the method. Andy et al. give a statistical analysis on consumption preference, consumption behaviors of tourists and crews, the number of people in Costa Rica of Latin America and finds that there was little difference in the contribution to local economy between cruise tourists and other tourists. Therefore, doubts were raised against the vigorous promotion of cruise tourism by local governments. Chase analyzes the impacts of tourism export industry on tourism destinations by using the theory of comparative advantages of tourism export, focused on the analysis for comparative advantages of an island economy in developing cruise tourism. Gregory Chase establishes a model to evaluate the impacts of cruise tourism on tourism destinations, and the effectiveness of Barbados's economic model for tourism destinations analysis.

Regarding the cruise tourism markets, Andriotis and Agiomirgianakis looks at the survey on tourist behaviors of cruise tourism in the Mediterranean Sea, which shows tourists in that region preferred such tourism activities as visiting historical relics, city sightseeing, shopping, excursions, visiting museums, dining, and swimming. Teye and Paris examine the activities of cruise tourists in the Caribbean Region, and results showed that tourists preferred such activities as tasting specialties, group touring, sightseeing, port shopping, entertainment, communication with local people, city sightseeing and others. Petrick et al. use decision-making variables to classify cruise tourists into two categories: complicated consumption decision-making tourists and quick decision- making tourists with a higher level of creditability. Brida analyze the questionnaire surveys on cruise tonrists in Cartagena in order to find out their consumption and expenditure structure, and results show significant difference in tourists' expenditure. Hung et al. look at the motivations of cruise tourism, and examines that the main motivations included cultural exploration, elusion and relaxation, recreation and entertainment, sightseeing, social motivation and the need for seeking new things. Hosany and Witham conduct research on the relationships between tourist experience and satisfaction or intention for recommendation. The results reveal that among the four dimensions of education, entertainment, beauty appreciation and elusion, beauty appreciation has the greatest impact on satisfaction and intention for recommendation of cruise tourists, followed by entertainment, education and elusion. Sanz Blas examines the relationships between satisfaction of cruise tourists for culture in tourism destinations and their intents of behaviors in the future. By comparing the differences between the cruise tourists from the U.K. and the U.S. and the tourists from Germany and Italy, the author concludes that the expectations to the trauel destinations have direct impact on tourist satisfaction, which in turn directly affected the tourists' decison-making on the next trip.

Regarding the environmental impact of cruise tourism, Butt looks at the management and treatment of cruise ship pollutants by cruise ports and the impact on ports. He proposes that emission reduction measures should be implemented to recycle pollutants. Jones explores the social and environmental impacts of the tendency towards giant cruise ships, and finds that large cruise ships have great influence on coral reefs. Relevent data reveals that sediment disturbance produced by the large-sized cruise ships is much more than natural water flow, and might cause significant adverse effect on corals in offshore areas. Caric argues that the change in cruise tourism usually reflects the transition of the external environment. Cruise ships would have major impacts on the main environment, and that such impact could be greater as the industry grow gradually.

2.2 Domestic Research

With respect to economic benefits of the cruise tourism industry, Wang et al. examines the economic output effect of China's cruise industry from the perspective of cruise industry chain. Miao et al. explore the promotive effect of tourist consumption on China's cruise industry from the perspectives of the demand for and the market supply of cruise tourism. Zhang et al. and Zhang et al. give an in-depth analysis of the features and problems in China's cruise tourism, and provides some policy suggestions on the development of cruise tourism in China. Liu et al. compare the development of cruise tourism in China, Japan and South Korea, discuss the problems in the industry chains, ports, tourism routes and others and put forward the ways to deal with then.

Regarding the consumption in the cruise tourism markets, Ye et al. recommend the cruise tourism products to satisfy tourists' demands in Shanghai market from the perspective of market demands. Sun et al. introduce a customer value spatial model into the marketing of cruise tourism and suggest strategies on the marketing of cruise tourism. Cai et al. (2010) analyze the Major industrial factors and competition potential of cruise tourism and establish a competition potential evaluation system for cruise tourism so as to evaluate the cruise tourism from the aspects of resource abundance, tourism operation performance, enterprise operation, financial and insurance services and traffic. And a factor analysis method was used to evaluate and analyze top 8 cruise tourism cities in China. Zhang et al. analyze the data on the international cruise tourism market, summariz the basic features of international cruise tourists and forecast the demand of China's cruise tourism market. Yang et al. conduct analysis on the features of China's cruise tourism market Jia et al. analyze the factors influencing the demand for cruise tourism, establish a prediction model for cruise tourism demand on the basis of BP neural network, conduct tests based on statistical data from the cruise market in the U.S. and predict the demand for cruise tourism in China. Zhang et al. establish a tourist consumption decision-making model for structural equation analysis of the data obtained, and the results show that motivations, opportunities and capabilities have positive effects on decision-making of cruise consumers. Sun et al. (2015) apply X-12-ARIMA and TRAMO/ SEATS seasonal adjustment models in the research on the data from the North American cruise markets, and the results show that the cruise market in North America continues to be expanded with more significant features.

Regarding port development, Wang et al. analyze the development status of port tourism resources and put forward the strategy for China's development of port tourism, i.e. the strategy for development of cruise tourism bases. Ye et al. conduct investigation by questionnaires and interviews, suggest to establish a standard system for home port services of cruise tourism, and put forward standardized development of cruise home ports from both national and local perspectives. Wu et al. establish a competitiveness evaluation system for international cruise ports, use the cloud model evaluation method to convert qualitative description into quantitative measurement, and conduct comparative competitiveness evaluation on the Port of Barcelona and Kotor in Europe, and the Port of Shanghai and Tianjin in China.

2.3 Summary

Literature reviews on domestic and overseas cruise industry development have revealed the following features and trends. In overseas, the focus of research is on the economic effects of cruise tourism and tourists' consumption behaviors both at home and abroad. In contrast, studies on cruise tourism in China focus on marketing and port development. Researchers think highly of the impacts of the travel patterns of cruise tourism in the industry. Another focus is on the economic impacts of cruise tourism Research methods are systematic, and quantitative research, instead of the qualitative, is a major method and to be think highly of by the researchers and experts Empirical analysis for quantitative research is highly valued, although qualitative research is still far from being sufficient. The research on cruise industry still needs to be conducted at a greater depth and breadth.

3 The Development of Cruise Market in China in the Past Decade

3.1 Continuous Scale-Up of China's Cruise Industry Over the Past Decade

Over the past decade, the supporting facilities for cruise industry have been improved. Infrastructure construction in cruise ports is in a steady progress. Custom clearance policies for cruise tourists have been improved; cruise port services have been promoted to higher levels; cruise industry chain has been extended from ports to inlands and the cruise culture has been cultivated at a quicker pace. The past decade witnesses the development of China's cruise industry for laying a solid foundation and keeping exploration. Supporting infrastructures in cruise ports, including tourist transport buildings and surrounding commercial facilities are being improved. And the industry also has been continuously exploring a set of standards for cruise port tourist services to gradually optimize the cruise tourist service procedures and increase the sense of satisfaction of tourists.

Regarding cruise routes, cruise ships originating in Chinese ports are mainly operating in China—South Korea, Sanya—Vietnam, cross-Straight and Hong Kong—Taiwan routes, mostly being 4–7 days' medium and short-term routes. Among these cruise routes, the routes between China and Japan/South Korea occupy a dominant position, accounting for over 98%.

With a decade's development, China's cruise tourism market keeps scaling up, with more market potential being explored and cruise tourism brand images being formed rapidly. This leads to enhanced recognition of tourists for cruise tourism. With the increased trips of cruise ships and tourist population at home ports, there is a significant decrease in both the number of cruise ships in visited ports and the number of tourists. Since 2014, China's cruise tourism has experienced explosive growth, resulting in a total of 466 navigation trips operated by ten major cruise ports in the country, increased by 15% on a year-on-year basis. The number of outbound and inbound cruise tourists reached 1,723,400, increased by 43% on a year-on-year basis, a strong momentum of growth. In 2015, the number of navigation trips operated by ten major cruise ports reached 629, and the number of outbound and inbound cruise tourists reached 2,480,500, making the country becoming the fourth largest tourist origin of cruise tourism in the world. Today, a consensus has been reached in the cruise industry: China is an emerging market in the world where cruise tourism has experienced the fastest growth, and China is expected to become the largest cruise market in the world in the future (Figs. 1 and 2).

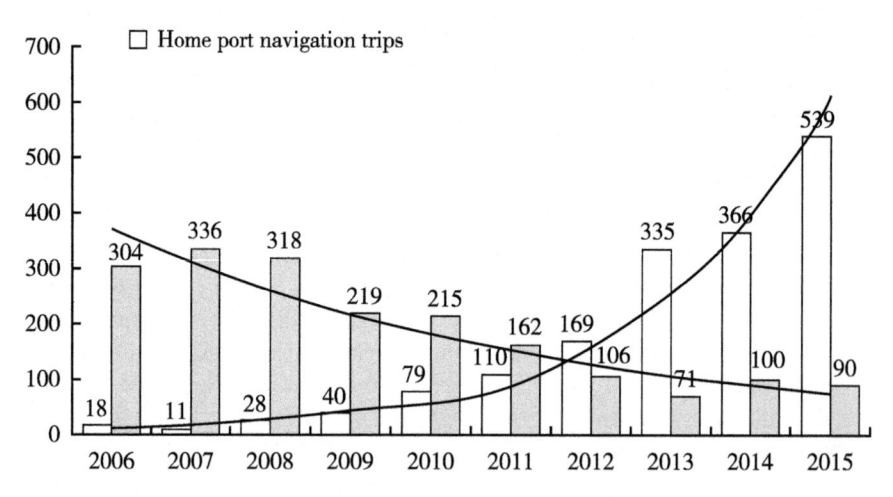

Fig. 1 Trips of cruise ships received in home ports and visited ports in China during 2006–2015. *Data source* Statistics of CCYIA, excluding data of Hong Kong and Taiwan

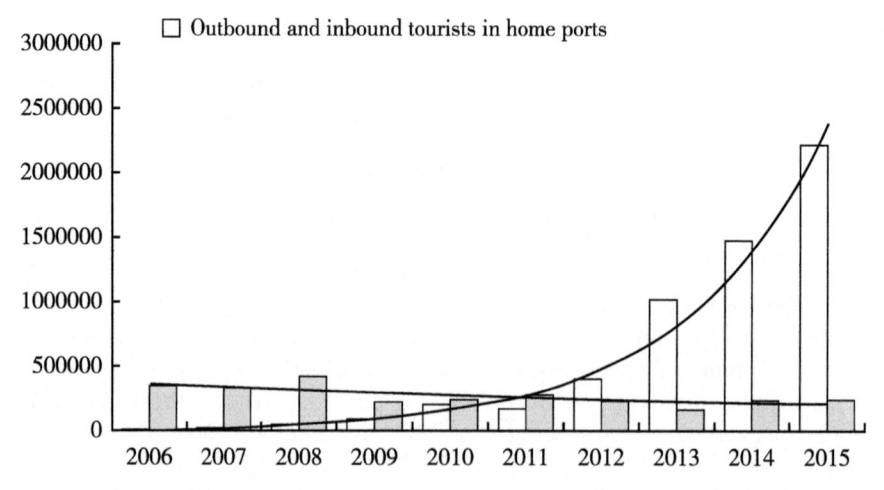

Fig. 2 Statistics of outbound and inbound tourists received at home ports and visited ports in China during 2006–2015 (unit: person-times). *Data source* Statistics of CCYIA, excluding data of Hong Kong and Taiwan

3.2 Improvement of Distribution of Cruise Ports in China Over the Decade

Over the decade, China has established five port clusters separately in the circum-Bohai-Sea area, Yangtze River Delta, regions on both sides of the Taiwan Strait, Pearl River Delta and southwest coastal regions. On this basis, regional cruise tourism markets are developed and centered on the cruise ports in Tianjin,

Shanghai, Xiamen, Shenzhen and Sanya. Each port cluster has its own locational advantages and highly developed traffic to form a tightly connected traffic network with surrounding tourist attractions. This achieves three-hour non-stop arrival at almost all tourism resources in these regions.

Shanghai's cruise industry develops the fastest among inland cities in China. In 2003, a development plan of cruise home ports was first proposed in Shanghai, and played an exemplary role in the country. At present, an international cruise combination home port has been built in Shanghai on the basis of Shanghai International Passenger Transport Center and Shanghai Wusongkou International Cruise Terminal, realizing the objective of "two terminals for one port". Attributed to its unique market and locational advantages, the Port of Shanghai became the largest cruise tourism hub in the Asia-Pacific Region and the eighth largest cruise home port in the world in 2015. Due to increasing imbalance between demand and supply, Phase 2 Project of Shanghai Wusongkou International Cruise Terminal was commenced in June 2015. When completed, the terminal can harbor four large-sized cruise ships at the same time.

Main works of Phase 2 of Tianjin Dongjiang Cruise Home Port was completed in early 2014. This port is capable of harboring a 220,000 t cruise ship at the maximum. The number of international luxury cruise ships received by the Port of Tianjin has increased year by year. This is attributed to the development of the port itself as well as its proximity to Beijing (the nearest port to Beijing), making it a cruise center in north China. Sanya Phoenix Island International Cruise Terminal, officially completed and open to navigation in 2006, is the first 80,000 t cruise terminal in China. Xiamen is among these ports that received international cruise ships the earliest in mainland China. In May 2014, the Port of Xiamen became the fourth international cruise home port in China.

3.3 Increased Input of Transport Capacity into Chinese Market by International Cruise Lines Over the Decade

Nowadays, international cruise lines are optimistic about the prospects of China's cruise market and have paid more attention to Chinese market, with more and more new luxury cruise ships being introduced into Chinese market. In 2006, there was only one home port cruise ship in operation. By the end of 2015, there had been as many as 12 home port cruise ships operating in China. Many cruise lines state that more luxury cruise ships of higher classes will be launched in Chinese market in the future. *Allegra*, the first home port cruise ship in China, only has a tonnage of 2850 t and carries 984 person. In 2016, China has 18 home port cruise ships, showing an obvious trend of scale-up of cruise ships introduced into Chinese market, including two cruise ships of over 160,000 t, six ships of 100,000–160,000 t, six ships of 70,000–100,000 t and four ships below 70,000 t. Royal Caribbean International has introduced its state-of-the-art cruise ship, *Quantum of*

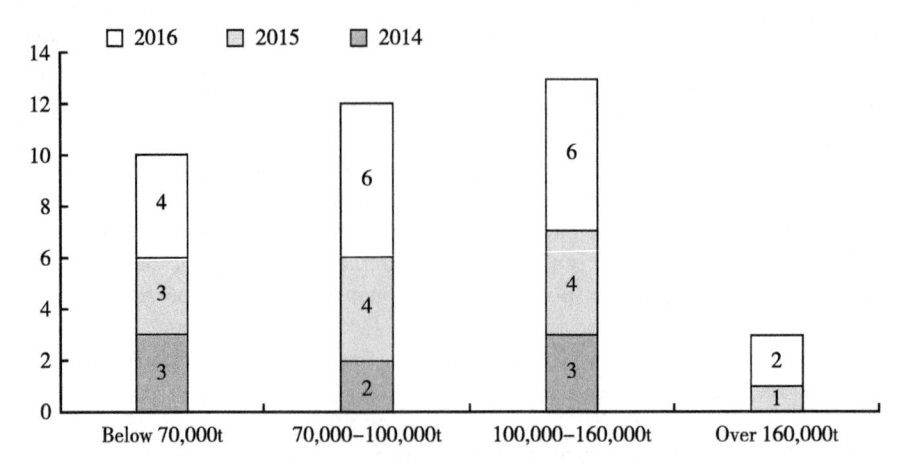

Fig. 3 Changes to transport capacity of cruise ship fleet in China during 2014–2016

the Seas, into Chinese market, and plans to introduced the *Ovation of the Seas*, one of its quantum class cruise ships, into Chinese market in 2016. By then, China will have 17 home port cruise ships in operation, forming an obviously larger fleet of cruise ships (Fig. 3).

3.4 Rapid Development of Online Cruise Tourism in China

According to the experience on the international cruise industry, it is essential to establish appropriate marketing channels. For cruise ticket sales, travel agencies have always been the most important channel. With greatly increased demand in the cruise market in China and continuous growth of the industry, travel agencies begin to energetically develop cruise business. In December 2012, a cruise channel went online at LY.com. In September 2014, ctrip. com initiated a cruise ticket reservation platform in China. With traditional outbound tourism impacted by cruise tourism, conventional travel agencies established cruise tourism departments one after another to cope with the competition between outbound tourism and cruise tourism. Travel agencies in tier-2 and tier-3 cities also began to step into the cruise market, aiming at the gradual increase of market size (Fig. 4).

Online cruise tourism platforms play an important role in cruise ticket sales. In 2015, total income of China's cruise market recorded CNY 4.53 billion, with the number of outbound tourists reaching 2,185,000. In the total income of China's cruise market, there is CNY 1.33 billion from online cruise reservation platforms, accounting for 29.3%, with an increase of 60.1% on a year-on-year basis. Among all online cruise reservation platforms, the trade amount from OTA accounts for 48.7% and that from self-owned online platforms of travel agencies accounts for 24.3%. OTA has become a platform experiencing the rapidest growth among all online platforms (Fig. 5).

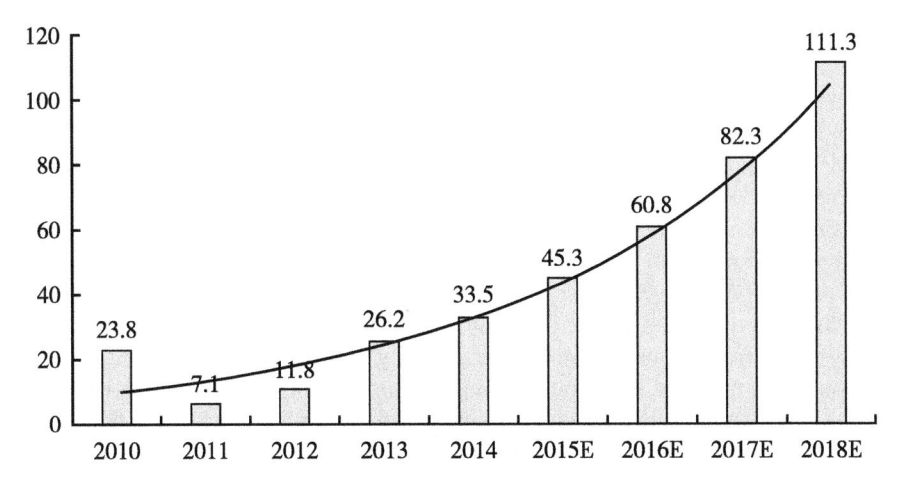

Fig. 4 Income of China's cruise market during 2010–2018 (unit: hundred million yuan). *Data source* Estimates by BigData-Research based on the data from CLIA

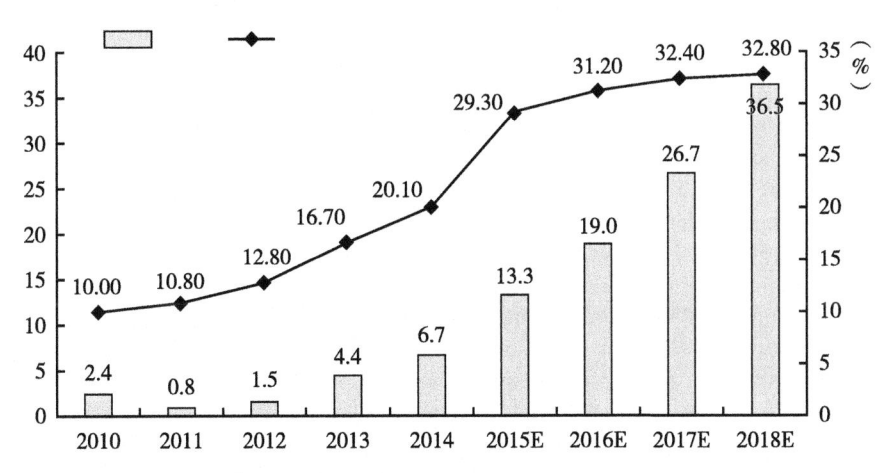

Fig. 5 Income of China's online cruise market and its percentage in total income during 2010–2018. *Data source* BigData-Research

3.5 Continuously Increasing Policy Support

For the better development of cruise tourism in China, central and local governments have introduced a package of preferential policies to establish an ideal environment for developing cruise tourism. To seize the opportunity for the development, National Development and Reform Commission (NDRC) issued the *Guidance for Promoting the Development of Cruise Industry in China* in June 2008. For the purpose of standardizing the development of cruise tourism, the

Ministry of Transport issued the *Guiding Opinions on Promoting the Continuous and Sound Development of Cruise Transport Industry in China* in March 2014. To advance coordinated development of ports, the Ministry of Transport issued the *Plan for Distribution of Cruise Terminals in Coastal Areas in China* in April 2015. In May 2015, the State Council issued the *Notification on Printing and Distributing "Made in China 2025"*, which requires breakthroughs be made in design and construction of luxury cruise ships. To extend China's cruise industry chain and have a more powerful say in the world,, the Ministry of Transport published the *Opinions on Further Promoting Tourism Investment and Consumption* issued by the State Council in Aug. 2015, requiring the establishment of domestic R&D, design, construction and autonomous supporting systems for large cruise ships and encouraging qualified domestic ship manufacturers to research, develop and construct large and medium-sized cruise ships. In September 2015, the Ministry of Industry and Information Technology of the People's Republic of China, NDRC and other four authorities jointly issued the *Implementation Opinions on Promoting the Development of the Tourism Equipment Manufacturing Industry*, in which five priorities are specified, with the top priority being "accelerating the realization of autonomous design and construction of cruise ships". To promote breakthroughs in the upstream industry chain, the 13th Five-Year Plan states that vigorous support should be provided for breakthroughs in core technologies (design, construction, etc.) of the cruise industry. The substantial support from central and local governments has provided ideal opportunities for the development of the cruise industry, and continuously enhanced the advantage and competitive edge of China's cruise tourism.

To eliminate the development bottlenecks as early as possible, promote innovations at the levels of systems, mechanisms, policies, regulations and standards, conduct reforms, seek breakthroughs with respect to industry, services and markets, enhance functionality of cruise home ports and explore new modes and ideas for the cruise industry, National Tourism Administration has successively approved experimental areas for cruise tourism in Shanghai, Tianjin, Qingdao and Shenzhen, playing a good demonstrative and leading role in promoting the development of China's cruise industry.

4 Analysis on the Issues in the Development of Cruise Industry in China in the Past Decade

4.1 Market Dominated by Foreign Cruise Ships and Slow Development of Domestic Cruise Ships

Currently, foreign cruise lines still have the initiative in China's cruise tourism market. Foreign cruise ships have occupied an absolutely dominant position in Chinese market, with over 95% of market share held by three international cruise giants: Royal Caribbean International, Carnival Cruise Lines and Genting Hong Kong Limited. Economic benefits brought by the cruise industry outflow, resulting in increasingly obvious contradiction between the rapid growth of the cruise industry and the limited economic contribution to China's development. Extremely high barriers to entry make it difficult for local cruise lines who own a single cruise ship to obtain scale merit based on their operational costs as entrants. The expansion of transport capacity develops very slow, and local cruise lines have little say in the market. Most Chinese cruise ships are second-hand ones that have been withdrawn from the mainstream European and American markets. On one hand, they can hardly compete with international cruise lines. On the other hand, services are still in the exploration stage, showing the lackness of competitiou. From the perspective of passenger sources, outbound tourists account for over 90% in the cruise market in China. The percentage of international passengers is lower, showing significant decrease in inbound cruise tourism. Local cruise tourism based on coastal tourism resources in China has not been fully developed, and the strong demand for domestic cruise tourism can hardly be satisfied. The problems of structural imbalance between demand and supply of cruise tourism are serious.

Regarding the development of local cruise ships, restrictions of policies on the development of the cruise industry need to be further relaxed to promote the growth of local cruise lines. For example, many companies having sufficient capitals are deterred by the policies in different fields including purchasing of cruise ships, taxes, cruise routes, scope of operation, staffing for cruise ships, etc. Such policies have restricted constitution and development of local cruise lines. A rapidly developing market requires continuous policy innovations. Therefore, there are still great limitations of the policies related to the cruise industry.

4.2 Missed Upstream and Downstream Segments of the Cruise Industry Chain and Limited Economic Contribution

At present, the most important pattern of cruise tourism for Chinese tourists is to take a foreign cruise ship for outbound tourism. China's cruise industry chain mainly focuses on port construction, operation, travel agencies. Ticket sales income from cruise tourism, consumptions on cruise ships and consumptions at tourism destinations outflow to foreign companies. In addition, China benefits less than expected from its rapidly growing cruise industry due to the underdeveloped cruise industry and insufficient supporting commercial facilities in the cruise economy. The strong driving and absorbing effects of the cruise economy have not been fully reflected. In China, coastal cities have made huge inputs in construction of cruise ports, which turned out to be poor in earning profits due to few profitable items. Currently, only Shanghai Wusongkou International Cruise Terminal gained profits.

In the upstream segment of the cruise industry chain, the core of cruise economy is cruise lines, with their economic added value accounting for over 50% in total cruise economy, followed by the cruise construction industry which has a percentage of 20%. At present, the cooperation between international cruise lines and domestic companies is still limited to purchasing ships from international cruise lines for operation or directly procuring a cruise brand for operation, not involving the design and construction of cruise ships or even more processes in the whole cruise industry chain. For a long time, the design and construction of cruise ships as well as the supporting industry chain are concentrated in Europe, led by European shipyards from Italy, Germany and France, with an annual production of merely 8 cruise ships. In China, vessel design, construction and maintenance, which are of high added values and require high technologies are still in blank field. Comparing with Japan, South Korea and other countries, China's cruise ship construction industry is far lagged behind in terms of both technology and autonomous innovation.

With respect to the service standards for cruise ports, it is essential to establish a standard service system to normalize the development of international cruise home ports and improve the quality of service of international cruise home ports. In China, the cruise industry is an emerging one with immature port management. Currently, there is still no normalized service standard system set up for such an emerging industry. Due to the lack of managerial experience on operating cruise ports, domestic cruise ports are still operated and managed at low levels. Therefore, there is an urgent need to gain in the experience on operation and management of cruise ports.

In addition, there are also no service standards for cruise tourism services. Travel agencies need to further improve their services and optimize the customer service procedures. All procedures relating to tourist reception are needed to be further optimized and upgraded.

4.3 Low Richness and Diversity in Cruise Routes

Over 90% of home port cruise ships departing from China take Japan and South Korea as their destinations. In terms of route distribution of cruise lines, they always prefer to arrange cruise routes lasting for 4–6 days. Japan and South Korea are the nearest destinations for routes starting from China. Tourists also expect to visit more places at the same price. With tourists requiring higher diversity in destinations, China's cruise tourism has become less attractive due to the simplicity of cruise routes, resulting in low rate of re-visiting of tourists. The current rapid growth is mainly attributed to strong demand. This is unfavorable for the trend of development of cruise tourism in the long run as it limits the sound and long-term development of the cruise market.

4.4 Slow Cultivation of Cruise Tourism Consumption Culture

Cruise tourism originated from Europe and America, and is an emerging mode of tourism in China. Despite the development over the decade, the cruise culture is still far from being mature. People have not obtained deep cognitions on cruise tourism. The research conducted on cruise tourism are rather superficial. Cruise tourism is still very strange to many people, who barely understand the nature of the industry.

Some tourists even have some misunderstanding of cruise tourism during their first trips and show certain reluctance to such a mode of tourism. With respect to price, most products of cruise tourism are at lump sum price, with most costs included in the price during the trip. Therefore, the price is rather high. Chinese customers are not used to purchase all the entire tourism products before their trip starts. Regarding the cognition of consuming groups, cruise tourism is believed to be a comfortable and relaxation experience, which is, according to traditional cognition of Chinese people, more suitable for the elderly instead of young people. In terms of cruise dining culture, Chinese tourists are not very adaptable to foreign food. When tourists arrive at their destinations, it is usually a hush for them to complete their visiting tours, and it is seemed that it's difficult to understand this for some tourists. In addition, some tourists are doubtful about the safety of cruise ships, and do not dare to take a cruise tourism.

4.5 Big Gap Between Demand and Supply of Cruise Tourism Talents

Strong talent reserve is always the guarantee for sound development of the cruise industry in its development process. Currently, China is in lack of professionals in the cruise industry. With the rising of China's cruise market, there are more demands for international cruise management and service talents. It is expected that there will be a big gap between demand and supply of the talents required for management, operation, technology, navigation, etc. in both upstream and downstream segments of the industry chain. This will result in increasingly growing contradictions between the rapid growth of cruise tourism and the lack of professionals. However, despite such a trend, there were still few majors related to cruise industry in colleges and universities in China. The situation has only been slightly mitigated in recent years. Such contradictions between demand and supply have, to some extent, increased the difficulties in improving the quality of service in cruise ports.

It is estimated that the gap in talents in cruise tourism will reach nearly 300,000 by 2020. Currently, the supply of professionals in cruise tourism is merely over 1000 people each year, leading to an extremely huge gap.

5 Forecasts and Analysis on Future Development of Cruise Industry in China

5.1 Continuous Expansion of the Cruise Tourism Consumption Market with an Obvious Trend Towards Demand Diversity

In recent years, the demand for outbound tourism of Chinese citizens has grown greatly. Since 2012, China has been the largest outbound tourism market in the world, with overseas average consumption per capita ranking the first in the world; China was the most important tourist source market with consumption attraction. In the next three decades, China will still face a prime period of rapid growth of tourism.

With the development of cruise tourism and the coming of the era of mass tourism, tourists are willing to participate in cruise tourism to experience the charm of the travel. More and more cruise tourism travel agencies are promoting the development of the cruise tourism market. People know cruise tourism better and improve their recognizations on cruise lines and ships. The cruise tourism culture will continue to grow and become mature in the country. There will be more abundant cruise tourism products along with the pace of innovation to satisfy diversified demands of tourists. The diversity of cruise tourism products will give Chinese tourists more options to spend their cruise holidays as well as maintain and promote the attraction of cruise tourism to tourists and customer loyalty. With increasing enthusiasm about cruise tourism, Chinese tourists will tend to experience such a mode of travel with special values.

5.2 Accelerated Port Integration with the Trend of Oligopoly Competition

The layout of distribution of cruise ports in China has been gradually improved, with more and more ports put into construction. Local governments have invested a great amount of funds into port construction. However, China is still exploring the way of operation and management of cruise ports, mostly borrowing the operation and management mode of cargo terminals for the moment. In terms of profits of cruise ports, berthing charge is still the main profitable item of a cruise port. The lack of diversified profitable items results in poor profitability of ports and prolonged recovery of the huge investment made into these ports. When cruise economy is developed, it is necessary to provide supporting facilities for the cruise industry based on cruise ports. Funds are required to construct public and commercial facilities, leading to a great demand for funds in later construction stage.

At present, China Merchants Group has initiated home port projects in Shenzhen, Xiamen, Qingdao and Tianjin, and holds the ownership of Shanghai Wusongkou International Cruise Terminal via Sinotrans & CSC Holdings Co., Ltd. China Communications Construction Co., Ltd. owns Sanya Phoenix Island International Cruise Terminal. Such large state-owned enterprises have strong financial power and can make continuous interference into cruise ports. Oligopoly competition will occur in cruise ports.

5.3 Accelerated Development and Increased Market Influence of Local Cruise Ships

The future of the cruise industry is promising. China is currently exploring its own path to develop local cruise ships. Chinese capitals have been invested into *SkySea Golden Era*, *Chinese Taishan* and Diamond Cruise's *Brilliant of the Seas*.

China has successively issued a package of policies to encourage innovations in design, construction, operation and management of cruise ships. China has plenty of cruise port resources supported by ideal maritime conditions of the East Sea, circum-Bohai-Sea region, Huanghai Sea and South China Sea. Abundant tourism resources are located in coastal areas in China. A solid foundation has been laid in the field of ship design and construction, and local tourism companies have developed strong power in this regard. Moreover, China is the largest tourist source in the world. All these factors create favorable conditions for developing local cruise ships.

In the future, with more emphasis laid on the cruise industry by Chinese government, Chinese cruise tourism fleet and onshore tourism destinations of Chinese characteristics are expected to be established, local cruise tourism routes will be developed, and domestic and overseas tourists would be attracted to coastal, island and maritime tours. China will exert much more influence on the cruise market.

5.4 Constantly Revised Laws, Regulations and Standards for the Cruise Industry

In promoting continuous development of the cruise industry, central and local governments have successively issued new policies to vigorously support the cruise industry, including some guidance policies. To better promote the development of the industry, more policies need to be issued and improved. Currently, China still borrows the experience on cargo ship management in its cruise-related policies. This is unfavorable to the development of the cruise industry. In the coming year, efforts will be made to issue and improve the policies related to cruise ship construction, purchase taxes, tourist customs clearance, visa-free transit, cruise tourism service standard, set-up of headquarters of foreign cruise lines, taxes and fees for in-operation cruise ships, cruise financial issues, offshore cruise development, cruise talents cultivation, etc. Breakthroughs will be made for existing policy restrictions to integrate with the development mode of the international cruise industry.

6 Recommendations on the Development of Cruise Industry in China

6.1 Operation and Management of Cruise Ports

As for port operation and management, policy support from the governments should be sought for and fully utilized. Experimental areas for China cruise tourism development has been set up in Shanghai and Tianjin to continuously innovate the mode of developing cruise tourism, explore new ideas for cruise tourism. This plays a demonstrative and leading role in the development of China's cruise industry. These experimental areas are set up as an example to achieve the shift from infrastructure construction to both infrastructure and system construction, and to continuously improve the cruise industry chain. The core should be cruise home ports, with emphasis laid on the integration of upstream and downstream segments of the industry chain as well as the cultivation of all-round service functions.

It is recommended to continue to implement policy innovations and carry out the centralized supervision over cruise tourism featured administrative localization and one-stop service, to provide convenient and highly efficient customs clearance services for cruise lines, agencies and ship enterprises, to continue to optimize customs clearance procedures, paperless customs clearance, integrated customs clearance and single-window customs clearance to reduce costs, and to improve capability and promote further improvement of customs clearance efficiency of ports.

It is also recommended to promote the development of the cruise port reception service standard system to continuously promote international and standardized services of cruise ports, to set up service management standards which include cruise port construction, safety management, emergency handling, supervision and inspection, to develop service quality standards which include query, tourist building space, and comfort and cleanliness of tourist buildings, to set up tourist service quality standards which include tourist landing, arriving, shopping, catering, suits and VIP lounge as well as port operation standards.

6.2 Operation and Management of Cruise Lines

Cruise tourism routes need to be enriched and cruise ships' departing ports should be diversified. More departing ports should be put into operation besides the main ports in Shanghai, Tianjin, Sanya and Xiamen, including Dalian, Qingdao, Yantai, Zhoushan, Shenzhen, Guangzhou, Beihai and Haikou. The diversity of departing ports will further promote diversified cruise routes, increase market coverage of cruise tourism and provide more opportunities for tourists. Entertainment, catering, services, etc. on board should be further diversified and innovated to provide better experience for tourists.

With respect to market access and openness, cruise consumption should be encouraged; and constitution and operation of cruise operation should be supported. Efforts should be made to develop the ship supply system, to support cruise ship R&D and construction, etc., to promote and implement visa-free policy for foreigners at cruise ports, to establish a more convenient and relaxed commercial tourism environment, to implement the policy of relaxing the duration of stay in visa-free transit of some national public employees from 72 to 144 h, and to extend the policy gradually from air ports to sea, land and air ports to realize linked visa-free transit policy at relevant transit ports.

With respect to the financial and credit support, funds should be raised for cruise development to promote the innovation of cruise financial services and products and to cultivate domestic funded cruise lines. In terms of vessel leasing, it is suggested to break the limit that any Chinese capital company shall not release vessels for operation until they have had at least one free vessel. It's recommended to encourage local cruise lines to lease cruise ships for operation. In experimental areas for China cruise tourism development, measures should be taken to facilitate direct cross-border investment made by enterprises and encourage enterprises in the experimental areas to make direct investment, with direct cross-border receipt and payment as well as exchange with the bank.

In terms of tax preferences and exemptions, domestic commodities supplied to overseas cruise ships in experimental areas for China cruise tourism development should enjoy export rebates as general trade. Commodities directly imported and transferred to overseas cruise ships should enjoy protective trade tax policies as general trade. The functions of bonded warehouse of cruise ships should be expanded, with public bonded logistics center established.

6.3 Improvement of Special Laws and Regulations on Cruise Tourism

As a new type of tourism in China, cruise tourism plays the roles of both transport and tourism. Due to the hysteresis of law, there are only a few laws and regulations prepared for cruise tourism in China. Sometimes there are even no regulations provided to stipulate certain tourism activities. Thus, it is urgent to improve the system of laws and regulations on cruise tourism. When cruise tourism events are handled, the most common practice is to follow the provisions in *Tourism Law* and *Maritime Law*. There is difference between cruise ticket contracts and ordinary transport contracts, and it is required to prepare special cruise ticket contracts to identify the obligations and rights of travel agencies, cruise lines and tourists so as to settle or avoid irrational occupation of cruise ships by tourists. There are still no special laws and regulations on the supply of cruise ships. As a result, the cruise ship supply industry can not play the role of speeding up the development of port economy.

Regarding environmental pollution control, current laws and regulations are too relaxed to actually limit the maritime environment pollution caused by cruise tourism. Poor effects are realized due to weak operability. China has established an environmental protection law system which includes *Environmental Protection Law*, *Marine Environment Protection Law* and Law *on Environment Impact Assessment*. However, there is still a big gap in the area of compensation for pollution and damages caused by cruise ships. Mostly, administrative penalties are adopted for responsibility identification and control of maritime pollution, and there is a lack of effective compensation and remedial mechanism for the loss of maritime environment. As a result, it is necessary to enact special laws and regulations on cruise environment control to ensure there are always legal bases for penalties and compensations.

6.4 Improvement of Talent Training Mode

Currently, the educational structure of cruise tourism is rather simple in China. Due to the strict requirements for cruise talents and their practical abilities, it is necessary to implement the guidelines of industry-university-research cooperation and combine theories with practices closely. With respect to cultivation of talents, it is required to establish a comprehensive educational system including undergraduate education, higher occupational education and trainings. Students with undergraduate education can be appointed with core posts due to their solid knowledge foundation, in-depth education in operation and management theories and strong abilities to learn and innovate. Talents with higher occupational education can understand basic theories of cruise tourism and practice the knowledge well with proper innovation, and they can be cultivated to become an important part in operation and management of cruise ships. Trainings are mainly carried out for in-service staff to increase their understardings on cruise tourism.

Cruise tourism requires highly internationalized talents. For this purpose, international schools are recommended. Besides academic education, it is also required to enhance the input into occupational education and training so as to continuously improve the education and training system for cruise tourism.

Chapter 5
Research on Asian Cruise Business Climate Index

Ling Qiu, Juan Luo, Linkai Qi and Hong Jiang

1 Introduction

In recent years, the world's cruise giants have drawn their attention to Asia, especially the Chinese market. With the eastward expansion of the cruise market, the growth rate of Asia and Oceania cruise tourists is far more than that of Europe and the United States. During 2010–2015, the number of cruise tourists in Europe and the United States has dropped to 75% from 80% of the global market share, while the Asia-Pacific region has increased to 7% from the original 1.2%, especially for China, where the growth rate has reached to 30%, much higher than the world average growth rate of 8%. In the next few years, the Asia-Pacific market will continue to show high growth. It is expected that Asian cruise visitors will reach 5 million by 2020. In addition, because the Asian population is close to 3 billion, which its cruise economy market penetration is only five ten thousandths; therefore, the Asian cruise industry has huge potential and growth space.

Ling Qiu—Research field: Cruise economy, Strategic management.

Linkai Qi—Research field: Cruise economy.

Hong Jiang—Research field: Cruise tourism, Industrial economy.

L. Qiu (✉) · J. Luo · L. Qi
School of Management, Shanghai University of Engineering Science, Shanghai, China
e-mail: qiuling100@163.com

L. Qiu
Shanghai International Cruise Business Institute, Shanghai, China

H. Jiang
Shanghai University of Finance and Economics, Shanghai, China

© Social Sciences Academic Press and Springer Nature Singapore Pte Ltd. 2018
H. Wang (ed.), *Report on China's Cruise Industry*,
https://doi.org/10.1007/978-981-10-8165-1_5

The cruise industry is not only a high growth industry but also a highly volatile industry. As the Asian cruise industry occupies an important position in the global cruise market, its volatility will affect the healthy development of the global cruise industry. Therefore, it is necessary to use a rigorous and scientific approach to analyze and evaluate the climate state of Asia cruise economy. In this paper, the Asian cruise ship business climate index system is constructed and the climate index is measured to describe the fluctuation of Asian cruise economy, and the mechanism of publishing Asian cruise business climate index is put forward to lay the foundation for improving the monitoring and alerting capacity of the cruise industry.

2 Literature Review

2.1 Research Status of Climate Index

2.1.1 Foreign Research

Based on cycle theory and index theory, the business climate analysis method has been one of the most commonly used macroeconomic analysis tool at home and abroad, and has been widely used in macroeconomic management sector. The first important practice in the study of the climate index was the Harvard index of general economic conditions published on the *Economic Statistics Evaluation* by Harvard Committee on Economic Research (established by Harvard University) in 1919, which uses the diffusion index method to predict The ups and downs of the climate fluctuation, becoming are one of the most important ways to study the climate index today. At the beginning of the 20th century, economists W. C. Mitchell and A. F. Burns made a greater contribution to the study of the business cycle. In 1913, Mitchell published the *Business Cycle* on the basis of Changji's research, which was considered a groundbreaking work in the study of business cycle fluctuations.[1] Since then, the study of the climate index has developed rapidly. Among them, the synthetic index method can not only predict the ups and downs of the climate fluctuation, but also can describe the degree of the fluctuation, which becomes the main method of the study of the climate index. Since the 1960s, due to the practical use of the government, the climate monitoring system with the climate index as the core has gradually become mature. In the 1970s and 1980s, the Organization for Economic Cooperation and Development (OECD), based on the concept of "growth cycle", used the climate analysis method to study the business cycle fluctuations of its member countries and developed the climate index of each member countries to determine their benchmark date of business cycle fluctuations. Nowadays, nearly 80 countries and regions use the climate index system to release information on

---

[1]Measuring Business Cycles, NBRE [M], 1947.

economic development. Among them, the NASDAQ index, ZEW business climate index and Nikkei stoke average are the relatively mature comprehensive index.

2.1.2 Domestic Research

Few studies are conducted on this field from the theoretical point of view, and most focus on the application analysis. The non-model based synthetic index method is a commonly used method. Since the 1980s, Chinese government agencies and scholars have been involved in the study and compilation of the relevant climate index. In 1987, the Jilin University Scientific Research Group, represented by Professor Dong Wenquan, cooperated with the former State Economic Commission for the first time to construct the diffusion index and the composite index of China's macroeconomic cycle.[2] Since then, China has shaped the Zhongjing index, entrepreneur confidence index, business climate index and Zhongjing industry climate index step by step. At the same time, our scholars are also involved in the study and compilation of the climate index. The earlier research includes the SW-type climate index[3] built by Chen and Gao (1994) for the macro economy. In 1997, Wang conducted a study on the business climate index with the method of diffusion index. At the beginning of the 21st century, Chinese scholars began to carry out empirical research in various fields with the method of climate index, and now empirical research covers iron and steel manufacturing, agriculture, tourism, real estate and some other fields.

2.2 Research Status of Climate Index in Tourism Industry

2.2.1 Foreign Research

Foreign research on the tourism climate index began in the 1980s. In 1985, Sheldon P. J. and Var T. tried to use the empirical method to predict the number of inbound tourists; in 1997, Wong K. F. began to use the business cycle to predict international travel demand; in the same year, Turner L. W., Kulendran N. and Fernando H. introduced the concept of leading economic indicator, and predicted the demand on the Australian tourism of tourists in Japanese, New Zealand, the United Kingdom and the United States with the synthetic index method. Foreign

---

[2]Dong Wenquan. Measurement, Analysis and Forecast of China's Business Cycle (1)—Existence and Determination of Business Cycle [J]. Journal of Social Sciences of Jilin University, 1987, 3 (1): 1–8.

[3]Chen Lei, Gao Tiemei. Using Stock-Watson-type Climate Index To Analyze And Forecast Macroeconomic Situation [J]. The Journal of Quantitative & Technical Economics, 1994, 5 (14): 53–59.

research on the tourism climate index is mostly focused on the forecast of tourism demand in various regions.

2.2.2 Domestic Research

In the early 21st century, Chinese scholars began to study the tourism climate index, and now the research object has been expanded to the tourism industry, travel agency industry, hotel industry, inbound tourism, heritage sites tourism and tourism in specific provinces and cities, etc. In the tourism industry, Ni and Dai (2007) took the lead in studying the climate index for China's tourism market, established the climate index system of China's tourism market, and used the data of 1985–2005 to carry out empirical research with the method of synthetic index[4]; Wang (2010) proposed a new tourism climate index system, and conducted empirical research according to the updated 1990–2009 data[5]. As for travel agency industry, Dai et al. (2007) established the travel agency climate index system, and conducted empirical research. With respect to the hotel industry, Dai Bin, Yan Xia, You Hao et al. established the hotel industry climate index and conducted empirical research. It can be seen that the research and development of the calculation and compilation method of the climate index are relatively mature. However, the empirical study on the tourism climate index is different. Different scholars choose different index systems according to their own methods, resulting in the different results in the calculation of the climate index.

2.2.3 A Review of Literature Researches on Tourism Climate

According to the domestic and foreign literature on the tourism climate index, we can see that on the one hand, foreign research on tourism climate index is superior to China, and the research on the concept of tourism climate index, theoretical system, method system and empirical research has been conducted in an earlier period, while China's research is based on the experience of foreign research; on the other hand, Chinese scholars' research on the relevant theory is relatively systematic, and the content of the empirical research is rich, which lay a good foundation for our study of tourism climate index. However, some shortcomings must also be pointed out: there is a logical disorder of the subordinate relationship between the research results on the choice of indicators; in the determination of weights, it is used to select main elements that are not mature to determine the weight, and when complex evaluation methods are used, there is no effective

---

[4]Ni Xiaoning, Dai Bin. Calculation and Analysis of Climate Index in China Tourism Market [J]. Journal of Beijing International Studies University (Tourism Edition), 2007 (11): 1–10.

[5]Wang Xinfeng. Empirical Research on China's Tourism Climate Index [J]. Statistical Education, 2011, (11): 55–60.

integration with these methods; with respect to the evaluation conclusion, the climate index concluded by the scholars are contradictory with each other and even contradictory in itself, thus there is no convincing conclusions.

3 Definition of Cruise Business Climate

3.1 Climate Index

The term climate originates from the business climate. It refers to the overall economic development trend. As a holistic description of the state of macroeconomic development, it is a business concept used to describe and analyze the vitality degree of economic activity. The climate index is the measure of the degree of climate. It is the monitoring result of the business cycle fluctuations. It not only describes the actual economic operation, indicates the expansion and contraction of the business, but also predicts the future development trend of the business. The climate index presents business fluctuations in the form of statistical indicators, which is intuitive and concise. The climate index can be divided into diffusion index and composite index according to the compilation method.

3.2 Cruise Business Climate Index

As a holistic description of the business development of cruise industry, cruise business climate is a business concept used to describe and analyze the vitality degree of cruise business. The cruise business climate index is the measure of the degree of cruise business, and the cruise economy fluctuates in the form of statistical index. The expression of the cruise business with the statistical index is the monitoring result of the cruise business cycle fluctuations. It not only describes the actual economic operation, indicates the expansion and contraction of the cruise business, but also predicts the future development trend of the cruise business. Based on the monthly or quarterly economic statistical data, the cruise business analysis is mainly used to analyze and determine the stage of cyclical fluctuations: expansion stage or contraction stage, peak point, valley point or turning point, and to find out the reasons for the change of climate state; meanwhile, to predict the future trend of cruise business climate and the emergence of the next peak or valley, and verify and evaluate the effectiveness of the implementation of cruise business policy, etc.

4 Establishment of a Climate Index Evaluation System for Asian Cruise Business

4.1 Indicator Selection Principle

Climate index analysis is based on correct selection of a climate index system, which comprises a range of highly sensitive and noticeable indicators that reflect the overall size, level and speed of the economy from different aspects. Primary principles for selecting proper indicators include: (1) Representativeness. Indicators for Asian cruise business should be representative of the development status of Asian cruise business as a whole. (2) Sensitivity. The indicators selected should be sensitive to economic fluctuations, as an "indicator" or an "alert". (3) Comprehensiveness. The indicators selected should be used to enable comprehensive analysis resulting in correct conclusions over various aspects of the economy. (4) Operability. The indicators selected should be supported by timely and easily accessible data. (5) Regularity and stability. Variables not belonging to the economic system, such as force majeure and social incidents, should be excluded. In addition, among collected statistical indicators, those of similar cycle and mild fluctuation should be selected. (6) Applicability Indicators should be generally applicable so that convenient conversion is possible given the diversified statistical standards of different countries.

4.2 Indicators Classification Methods

Based on the time difference between changes of business indicators and the overall economy, business climate indicators are classified into leading indicators, coincident indicators, and lagging indicators, which are therefore determined as the Level 2 indicators of the development climate index system for cruise business.

4.2.1 Leading Indicators

By comparison with business climate of cruise industry, these indicators fluctuate in advance. That is, they touch the bottom or reach the peak earlier than the fluctuation cycle of business climate of cruise industry, and hence represent some symptoms of the business climate in the next period to come. Leading indicators have been proposed to identify where the turning points are in the climate fluctuation cycle, and to predict the amplitude of such climate fluctuation.

4.2.2 Coincident Indicators

Coincident indicators are used in this paper to represent characteristics of business climate fluctuations of the cruise industry. While turning roughly at the same time with regional economic cycles, coincident indicators demonstrate the status quo of regional cruise industry instead of predicting the trend in the future. Study on the change of such indicators provides a clear business climate of the cruise industry.

4.2.3 Lagging Indicators

Lagging indicators fluctuate only after the actual cycle of business climate. In other words, the bottom and peak of the lagging indicators come some time later than that of the business climate of the cruise industry. Lagging indicators are proposed to help verify the certainty of business trend represented by the leading indicators.

4.3 Establishment of the Evaluation Index System

According to several indicator selection methods, subjective or objective, such as textual analysis, literature research, expert scoring and consultation, and based on the screening principles of representativeness, operability and applicability, the business climate index evaluation system for Asian cruise industry has been finalized (see Table 1). It primarily includes 3 Level 2 indicators (leading/ coincident/lagging indicators) and 19 Level 3 indicators:

4.4 Determining Weights of Evaluation Indicators

Weight assignment of the evaluation indicators can be subjective and objective. In subjective assignment, a qualitative analysis method, the weight of each indicator is predetermined based on subjective experience or expert assessment. Two shortcomings of current subjective weight assignment methods are as follows: (a) a particular factor can be estimated too high or too low, hindering the true situation of targets evaluated from being fully represented; and (b) the weight, generally speaking, will not be changed once determined, while the inspection system selected by evaluation actors is flexible, further downgrading the credibility and accuracy of the comprehensive evaluation. To avoid the aforesaid shortcomings, Coefficient of Variation, a more objective method, is adopted in the paper for weight assignment.

First, an evaluation index system is established to include i indicators ($i = 1, 2, \ldots, n$) for evaluation of j objects ($j = 1, 2, \ldots, n$); Second, calculating average values, the standard deviation and variable coefficient of all indicators. Average

Table 1 Business climate index evaluation system for Asian cruise industry

Level 1 indicators	Level 2 indicators	Level 3 indicators	Unit
Cruise business climate index of Asia	Leading indicators	GDP	Trillion yuan
		Per capita disposable income	10,000 yuan
		Economic growth rate	%
		Total number of tourists received	10,000 passengers
	Coincident indicators	Cruise repair and maintenance base	Nr.
		Number of cruise ships manufactured	Nr.
		Cruise supply income	10,000 yuan
		Total number of cruise tourists received	10,000 passengers
		Number of cruise tourists received at home ports	10,000 passengers
		Total number of cruise ships received	Ships
		Number of cruise ships received at home ports	Ships
		Number of cruise ships at home ports	Nr.
		Average tonnage of cruise ships at home ports	10,000 ton
		Total number of cruise lines	Nr.
		Per capita consumption of inbound cruise tourists	1000 yuan
		Per capita consumption of outbound cruise tourists	1000 yuan
		Number of berths at the cruise terminal	Nr.
	Lagging indicators	Number of onshore workers of cruise terminals and cruise lines	Nr.
		Average salary of onshore workers of cruise terminals and cruise lines	1000 yuan

value of ith indicator is calculated by $\bar{x} = \frac{\sum_{j-1}^{m} x_{ij}}{m}$, standard deviation of ith indicator is calculated by $\sigma_i = \sqrt{\frac{\sum_{j-1}^{m}(x_{ij}-\bar{x}_1)^2}{m}}$, and the coefficient of variation of each indicator is calculated by $V_i = \frac{\sigma_i}{\bar{x}_1}$. Third, determining weights W_i of indicators according to the formula $W_i = \frac{V_i}{\sum_{j-1}^{n} V_j}$ and SPSS 22.0 software, the final weights of indicators are calculated and shown in Table 2.

Table 2 Weights of Asian cruise business climate indicators

Level 1 indicators	Level 2 indicators	Level 3 indicators	Unit	Weight
Cruise business climate index	Leading indicators	GDP	Trillion yuan	0.054
		Per capita disposable income	10,000 yuan	0.054
		Economic growth rate	%	0.076
		Total number of tourists received	10,000 passengers	0.065
	Coincident indicators	Total number of cruise tourists received	10,000 passengers	0.069
		Number of cruise tourists received at home ports	10,000 passengers	0.066
		Total number of cruise ships received	Ships	0.076
		Number of cruise ships received at home ports	Ships	0.062
		Number of cruise ships at home ports	Nr.	0.065
		Average tonnage of cruise ships at home ports	10,000 ton	0.047
		Total number of cruise lines	Nr.	0.058
		Per capita consumption of inbound cruise tourists	1000 yuan	0.076
		Per capita consumption of outbound cruise tourists	1000 yuan	0.060
		Number of berths at the cruise terminal	Nr.	0.058
	Lagging indicators	Number of onshore workers of cruise terminals and cruise lines	Nr.	0.062
		Average salary of onshore workers of cruise terminals and cruise lines	1000 yuan	0.052

5 Research on Determination of Cruise Business Climate Index of Asia

5.1 Methods for Determining Cruise Business Climate Index of Asia

Composite index (CI) method—In this method, some climate-sensitive indicators are selected from key statistic indicators of macroeconomic activities, and change rates of these selected indicators are combined to observe and monitor the intensity of business climate change. The composite index is used not only to predict where the turning points of business cycles will occur in the fluctuation, but also to somehow reflect the amplitude of such fluctuation. And this is why the composite index is used in this paper for determining cruise business climate index of Asia.

5.1.1 Obtaining and Standardizing Symmetric Change Rate of Indicators

Assume that $Y_{ij}(t)$ represents the ith indicator of the jth indicator set, where $j = 1, 2, 3$ and represents the leading, coincident and lagging indicator sets respectively, while i refers to the sequence of indicators ($i = 1, 2, \ldots, k_j$); k_j is the total number of indicators in the jth set. First, symmetric change rate $C_{ij}(t)$ of $Y_{ij}(t)$ should be obtained:

$$C_{ij}(t) = 200 \times \frac{Y_{ij}(t) - Y_{ij}(t-1)}{Y_{ij}(t) + Y_{ij}(t-1)}, \quad t = 2, 3, \ldots, n$$

When the component indicators $Y_{ij}(t)$ contain zero or negative values, or the indicators form a ratio sequence, take their first-order differences:

$$C_{ij}(t) = Y_{ij}(t) - Y_{ij}(t-1), \quad t = 2, 3, \ldots, n$$

Standardize $C_{ij}(t)$ of indicators so that the average absolute value equals 1 to prevent large deviation. Obtain the standardization factor A_{ij}:

$$A_{ij} = \sum_{t=2}^{n} \frac{|C_{ij}(t)|}{n-1}$$

Standardize $C_{ij}(t)$ with A_{ij} to obtain standardized change rate $S_{ij}(t)$:

$$S_{ij}(t) = \frac{C_{ij}(t)}{A_{ij}}, \quad t = 2, 3, \ldots, n$$

5.1.2 Obtaining Standardized Average Change Rate

(1) Assume $R_j(t)$ as the average change rate and calculate $R_j(t)$ of leading/coincident/lagging indicator sets respectively:

$$R_j(t) = \frac{\sum_{i=1}^{K_j} S_{ij}(t) \cdot W_{ij}}{\sum_{i=1}^{K_j} W_{ij}}, \quad j = 1, 2, 3; \ t = 2, 3, \ldots, n$$

W_{ij} refers to the weight of ith indicator of the jth set and since equivalent weight is used in the model, $W_{ij} = 1$.

5 Research on Asian Cruise Business Climate Index

(2) Calculate standardization factor of indicator F_j:

$$F_j = \left[\sum_{t=2}^{n}|R_j(t)|/(n-1)\right] \bigg/ \left[\sum_{t=2}^{n}|R_2(t)|/(n-1)\right], \quad j=i,2,3$$

Note: $F_2 = 1$.

(3) Calculate standardized average change rate $V_j(t)$:

$$V_j(t) = R_j(t)/F_j, \quad t = 2, 3, \ldots, n$$

For the purpose of using the leading/coincident/lagging indicators to form a system with consolidated structure, the change rate of leading and lagging indicators should be averaged first for adjustment based on the amplitude of unified indicator average change rate.

5.1.3 Obtaining Initial Composite Index $I_j(t)$

If $I_j(l) = 100$, then

$$I_j(t) = I_j(t-1) \times \frac{200 + V_j(t)}{200 - V_j(t)}, \quad j = 1, 2, 3; \; t = 2, 3, \ldots, n$$

5.1.4 Trend Adjustment

(1) Compound interest formula is used to obtain the average growth of each sequence of the coincident indicator set:

$$ri = \left(\sqrt[mi]{C_{L_i}/C_{I_i}} - 1\right) \times 100, \quad i = 1, 2, \ldots k_2$$

$$C_{I_i} = \left(\sum_{t \in \text{the first cycle}} Y_i(t)\right) \bigg/ m_{I_i}$$

$$C_{L_i} = \left(\sum_{t \in \text{the last cycle}} Y_i(t)\right) \bigg/ m_{I_i}$$

where C_{I_i} refers to the first circulated average value of ith indicator in the coincident indicator set; C_{L_i} refers to the last circulated average value of the ith indicator in the coincident indicator set; m_{I_i} refers to the month of first circulation of ith indicator in the coincident indicator set; m_{L_i} refers to the month of last circulation of ith indicator in the coincident indicator set; k_2 refers to number of coincident indicators; mi refers to months between the centers of first and the last circulation.

Designate the average growth of the coincident indicator set as G_r, i.e. the target trend:

$$G_r = \left(\sum_{i=1}^{k_2} r_i\right)/k_2$$

(2) The compound interest formula is used to calculate the average growth of initial composite index $I_j(t)$ of leading/coincident/lagging indicators ($j = $ S1, 2, 3) r'_j:

$$r'_j = \left(\sqrt[mj]{C_{L_j}/C_{I_j}} - 1\right) \times 100$$

$$C_{I_j} = \left(\sum_{t \in \text{ the first cycle}} I_j(t)\right)/m_{I_j}$$

$$C_{L_j} = \left(\sum_{t \in \text{ the last cycle}} I_j(t)\right)/m_{L_j}$$

Trend adjustment:

$$V'_j(t) = V_j(t) + (G_r - r'_j) \quad j = 1, 2, 3, \ t = 2, 3, \ldots, n$$

5.1.5 Calculation of Composite Index

If $I'_j(1) = 100$, then:

$$I'_j(t) = I'_j(t-1) \times \frac{200 + V'_j(t)}{200 - V'_j(t)}, \quad j = 1, 2, 3; \ t = 2, 3, \ldots, n$$

Composite index for base year of 100 is generated:

$$CI_j(t) = \left(I'_j(t)/\overline{I'_j}\right) \times 100$$

where $\overline{I'_j}$ is the average value of $I'_j(t)$ in the base year.

5.2 Data Collection

So many challenges have been confronted in data collection and statistics of the cruise industry since it connects with numerous sectors and it's hard to separate the data of cruise industry from others. Therefore, for a new attempt like measuring business climate index of cruise industry, lack of data sources is inevitable.

Statistics for the business climate index of cruise industry in Asia and China obtain the data from two sources: (a) official websites and publications such as China statistical yearbook, National Economic and Social Development Statistical Bulletin, CCYIA official website, CLIA official website, China Cruise Development Report, Green Book on Cruise Industry, annual reports of cruise terminal companies and annual reports of renowned domestic and international cruise lines; (b) investigation interviews with directors of Japanese and South Korean national tourism authorities, cruise terminals, Chinese and foreign cruise lines and cruise travel agencies.

5.3 Measurement of Business Climate Index of China's Cruise Industry

As the prerequisite of comprehensive evaluation, nondimensionalization of indicators is necessary as it enables the comparison of indicators with various attributes at the same level. Extreme difference method is applied in nondimensionalization of the data. Since the composite index method is applied in this paper, we first calculate the symmetric change rate of indicators and adjust it on a seasonal basis to obtain the average change rate of leading/coincident/lagging indicator sets and use it to calculate standardization factor. We take the year 2011 as the base year and set its value to 100 to determine the composite business climate index of China's cruise industry from 2011 to 2017 (see Table 3; Fig. 1), and have plotted the diagram of composite index, as shown in Fig. 2.

Table 3 Composite business climate index of China's cruise industry

Year	Leading indicator	Coincident indicator	Lagging indicator	Comprehensive composite indicator
2011	100.00	100.00	100.00	100.00
2012	100.25	100.34	100.47	100.29
2013	101.10	101.17	101.11	101.15
2014	101.31	101.41	101.28	101.37
2015	101.87	101.93	101.76	101.99
2016	102.31	102.16	101.97	102.27
2017	102.71	102.59	102.38	102.53

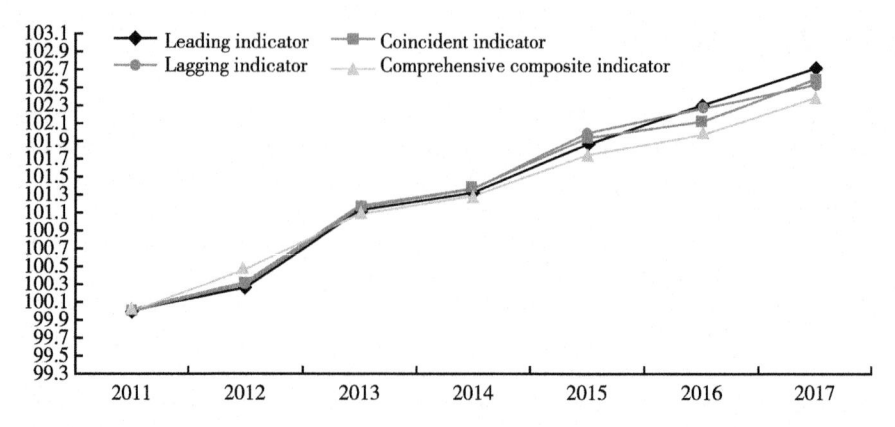

Fig. 1 Composite business climate index of China's cruise industry

Fig. 2 Business climate index of China's cruise industry

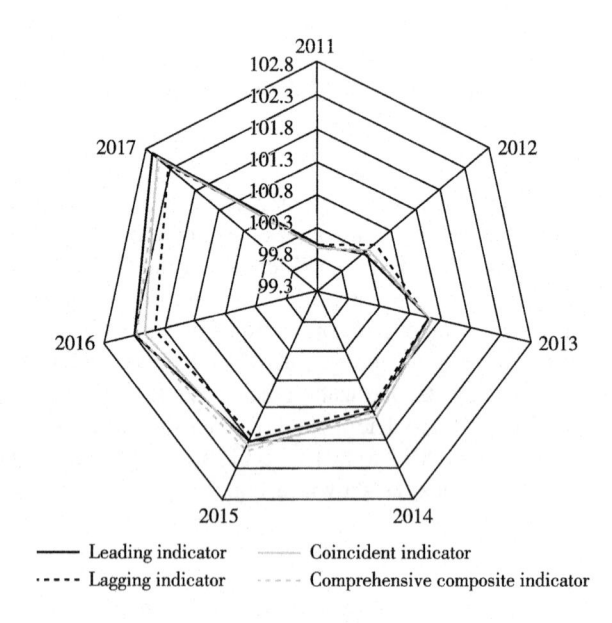

According to the figures and tables above, the leading indicator is more accurate than the coincident indicator in terms of predicting occurrence of peaks; and compared with actual occurrence of bottoms, the lagging indicator generally lags behind a bit and, while being mild and not very ideal, it lends some support to the actual business climate. From the beginning of 2011 to the end of 2017, the business climate index of China's cruise industry showed slight fluctuation, and the coincident indicator and comprehensive indicator tend to be concurrent with each other.

5.4 Measurement of Business Climate Index of Asian Cruise Industry

As the prerequisite of comprehensive evaluation, nondimensionalization of indicators is necessary as it enables the comparison of indicators with various attributes at the same level. Extreme difference method is applied in nondimensionalization of the data. Since the composite index method is applied in this paper, we first calculate the symmetric change rate of indicators and adjust it on a seasonal basis to obtain the average change rate of leading/coincident/lagging indicator sets and use it to calculate standardization factor. We take the year 2011 as the base year and set its value to 100 to determine the composite business climate index of Asia's cruise industry from 2011 to 2017 (see Table 4; Fig. 3), and have plotted the diagram of composite index, as shown in Fig. 4.

It is noticeable, according to the figures and tables above, that the leading indicator is more accurate than the coincident indicator in terms of predicting occurrence of peaks; and compared with actual occurrence of bottoms, the lagging

Table 4 Composite business climate index of Asia's cruise industry

Year	Leading indicator	Coincident indicator	Lagging indicator	Comprehensive composite indicator
2011	100.00	100.00	100.00	100.00
2012	100.21	100.29	100.41	100.19
2013	101.11	101.09	101.08	101.06
2014	101.24	101.35	101.21	101.3
2015	101.68	101.91	101.65	101.83
2016	102.08	102.11	101.79	102.07
2017	102.53	102.31	102.07	102.37

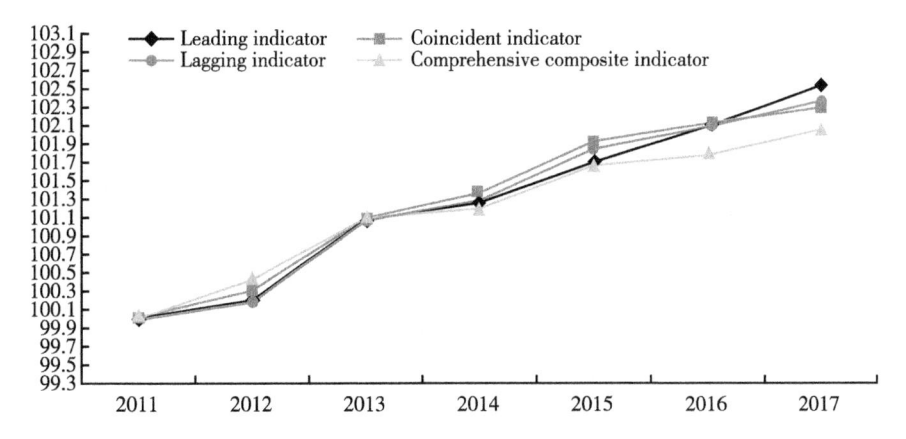

Fig. 3 Composite business climate index of Asia's cruise industry

indicator generally lags behind a bit and, while being mild and not very ideal, it lends some support to the actual business climate. From the beginning of 2011 to the end of 2015, the business climate index of Asia's cruise industry showed slight fluctuation, and the coincident indicator and comprehensive indicator tend to be concurrent with each other.

6 Establishment of a Publication Mechanism of Business Climate Index of Asia's Cruise Industry

6.1 Principles for Establishment of a Publication Mechanism of Business Climate Index of Asia's Cruise Industry

To establish a publication mechanism of business climate index of Asia's cruise industry, we should first identify the fundamental principles and ideas for the purpose. We should prioritize the numerous challenges confronting the cruise industry in establishing and improving the mechanism, follow the rules and characteristics of information dissemination, respect the public's right to know and the right of media coverage, and standardize and supervise the mechanism based on laws, so that an early-warning information publication system can be established to enable positive interaction among Shanghai International Cruise Business Institute (SICBI), the public and the press.

6.1.1 Principle of Openness

Openness of information should be a primary principle. The aim of building an all-media government warning publication mechanism is to maximize free circulation of information, ensure the public's right to know, eliminate negative impact of the early-warning, and therefore ensure stable social order.

6.1.2 Principle of Truthfulness

The information published should be true, accurate and complete. Truthfulness is a primary principle for information publication. This requires the publishers to ensure that the information published agrees with the objective reality and supported by sufficient and accurate data sources.

Fig. 4 Business climate index of Asia's cruise industry

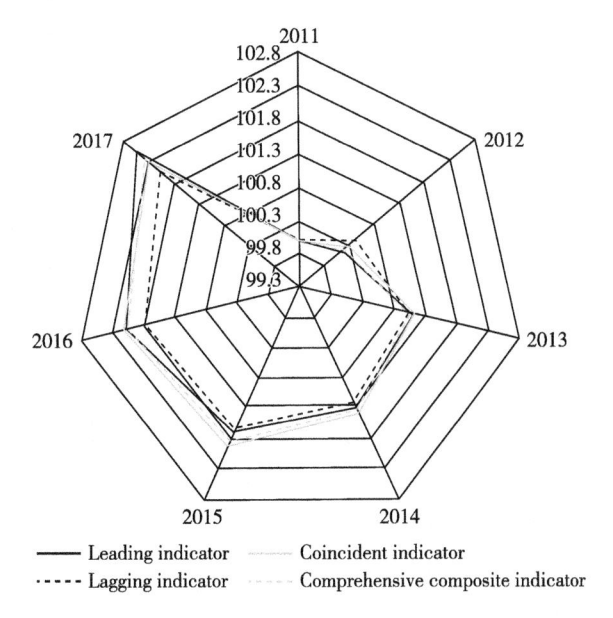

——— Leading indicator   ——— Coincident indicator
- - - - Lagging indicator   - - - - Comprehensive composite indicator

6.1.3 Principle of Regularity

SICBI regularly collects monthly, quarterly, or annual statistical reports on cruise industry and has made cruise statistics its routine work. It also publishes its own monthly, quarterly, or annual statistical reports and release monthly, quarterly, and annual data of cruise business climate indexes.

6.2 Building a Publication Mechanism of Climate Index of Asia's Cruise Industry

6.2.1 Establishing a Management Mechanism for Publication of Business Climate Index of Asia's Cruise Industry

The release of business climate index of Asia's cruise industry is important because the index reflects the development state of Asia's cruise industry and predicts its development trend, and thus provides significant reference for the development of cruise industry both in Asia and China. SICBI is committed to building an international, open and high-level decision-making advice and research platform focusing on policy-making theories, innovative ideas and development strategies. It has facilitated broad and diversified academic exchange in many ways, including releasing topics to the world, holding summit forums of China's cruise industry, publishing reports on China's cruise industry (Green Book on Cruise Industry) on a yearly basis. A series research results have provided policy-making basis and intellectual support to the governments and cruise lines for developing cruise policies and strategic plans. Therefore, SICBI is recommended to establish a special research group of cruise business climate index to improve the evaluation system of business climate index of cruise industry in Asia and China, and release the climate index on a regular basis.

6.2.2 Developing Statistical System of Cruise Industry

Shanghai Municipal Statistics Bureau is recommended to develop a statistical system of cruise industry. It should: (1) define cruise registration process, report procedure, statistical document submittal rules, manning scheme, and investigation methods. By referring to the tourism statistical system, the statistical reports of cruise industry can be classified into four categories: grassroots statistical report, department statistical report, specialty statistical report and tourism sample survey questionnaire; (2) appeal to the Ministry of Transport, National Bureau of Statistics, Ministry of Public Security and other national authorities for support to expedite the issuance of *National Cruise Ship Operation Statistic System* and *Measures for Management of Cruise Ship Operation Statistics* in order to identify the obligations and responsibilities of competent authorities; (3) improve statistical means for the cruise industry by developing cruise statistic modules in current terminal/port statistic systems, using information technology to optimize the cruise statistic and increasing the statistical efficiency; (4) study and establish an online registration system for statistics of cruise industry. The online registration system should be provided with access ports of different permission level to enable statistic staff of enterprises to enter and update the data online. This system simplifies the work process between administrations and enterprises; and (5) develop a comprehensive survey method. Based on general investigation, the method should take sample surveys as its major form, supported with full-scale surveys, typical surveys, key-point surveys, and scientific estimation.

6.2.3 Establishing an Information Collection System for Business Climate Index of Cruise Industry

Information collection and sharing are important to the cruise climate index. It is recommended that Shanghai municipal government should lead the collection information of cruise industry, with major responsibilities divided to Shanghai Municipal Statistics Bureau, Shanghai Municipal Tourism Administration, Shanghai Municipal Transportation Commission, Shanghai Municipal Office of Port Service, Baoshan District Government, Hongkou District Government, cruise terminal companies and SICBI, and supported by other government agencies or organizations relate to the cruise industry, such as Shanghai General Station of Immigration Inspection, Shanghai Customs, cruise lines and cruise travel agencies. Each such organization should regularly provide monthly, quarterly, or annual statistical reports on cruise industry and take the cruise statistics as their routine work. They should also publish monthly, quarterly, or annual statistical reports, and SICBI should release monthly, quarterly, and annual data of cruise business climate indexes.

6.2.4 Establishing a System to Support Statistic of Cruise Industry

The statistic of cruise industry is closely related to the research of cruise business climate index. A support system is required to manage and supervise the statistic process of cruise industry. Major organizations responsible for cruise statistics at all levels should establish data quality review divisions to evaluate data quality by applying modern statistic methods such as sample survey, logical evaluation, comparison analysis, and expert demonstration, based on computer technology, and using vertical extension and horizontal association of data. Problems, if any, should be tracked and verified timely. Practical quality control and evaluation methods should be developed based on circumspect studies on evolution characteristics and trend of statistical indicators of local cruise industry. A quality control system for cruise statistical data should also be established to control errors within a reasonable scope.

To reflect business climate of Asia's cruise industry more scientifically and accurately, the research team of business climate index of Asia's cruise industry will keep upgrading the cruise business climate evaluation indicator system and indicator weight assignment method based on the development status of Asia's cruise industry, establishing cruise business information statistic mechanism, gradually exploring statistic channels, and updating the business climate index regularly, so as to provide references for monitoring and early-warning of cruise industry both in China and Asia.

Chapter 6
Research on Cruise Industry in China: Current Issues and Prospects

Xiaodong Sun and Rongxin Ni

1 Introduction

Dating from the 1960s and with the rapid development of over 50 years, the modern cruise industry has developed into one of the fast-growing industries with the most remarkable economic benefits in global tourism and hospitality industry, known as a "golden industry on the golden waterway". In recent years, with the gradual saturation of international cruise market and the eastward-moving trend of the center of cruise market, China has become a strategic emerging market which the international cruise lines have been competing for. A number of international cruise lines, including Costa Crociere, Royal Caribbean Cruises, Princess Cruises, MSC Cruises, Star Cruises, Norwegian Cruise Line and Dream Cruises, have entered the cruise market in Mainland China.

In 2016, the number of cruise tourists from Mainland China was over 2,000,000, accounting for about 10% of total of the international cruise market. The huge market of cruise tourist source, intensive policy support and continuous improvement of recognition on cruise products will make cruise tourism become a significant new form of tourism industry in China. The development of the cruise industry also requires guidance of the academic circle on theories and experience. At present, the cruise industry in China is in a rapid expansion; however, corresponding academic research has not been developed in balance with such an industry expansion. It can even be said to be still lagging behind the development stage of the cruise industry. In order to enable the researchers to get a thorough understanding on the research situation in the academic circle of the cruise industry in China, relevant literature in over ten years is analyzed and reviewed in this paper, so as to better facilitate the research on the cruise industry in China.

X. Sun (✉) · R. Ni
East China Normal University, Shanghai, China
e-mail: xdsun@bs.ecnu.edu.cn

© Social Sciences Academic Press and Springer Nature Singapore Pte Ltd. 2018
H. Wang (ed.), *Report on China's Cruise Industry*,
https://doi.org/10.1007/978-981-10-8165-1_6

2 Research Overview

In respect of literature collection, with CNKI (www.cnki.net) as the database and "cruise" as the keyword, only over 100 journal articles are searched out. However, when cruise is searched among the titles and abstracts, more than 1100 journal articles are found, including over 160 articles in core journals. In terms of the quantity of published articles, the number of achievements in research on the cruise industry in China has grown rapidly since 2006, during which the research literature between 2006 and 2017 accounted for more than 97% of the total achievements; especially since 2012, such number has increased by leaps and bounds, with more than 800 papers published in just five years from 2013 to 2017, accounting for about 70% of the total, as shown in Fig. 1. In terms of publication journals, most of the papers are published in general journals, over 160 achievements in core journals and only less than 30 papers in the journals listed in Chinese Science Citation Database (CSCD) and Chinese Social Sciences Citation Index (CSSCI), among which over two thirds of the achievements were published after 2010. In this paper, the significant research achievements published in core journals are mainly analyzed, rather than simply review, in generalities, of the achievements in research on the cruise industry in China.

In order to attract the attention of Chinese scholars to research on cruise tourism, Zhang and Kou et al. (2012) reviewed the research on overseas cruise tourism markets and proposed research directions worthy of study in the future. Sun and Feng (2012) summarized and reviewed the achievements in research on the cruise industry in China from macro and micro perspectives respectively. The authors pointed out that the existing research contents in domestic literature are extensive and the research perspectives are not specific enough; it was suggested that research in the future be made from the perspectives of industrial cluster, stakeholder, cruise tourism market and the psychology and behavior of tourists, etc. In addition, Zhang

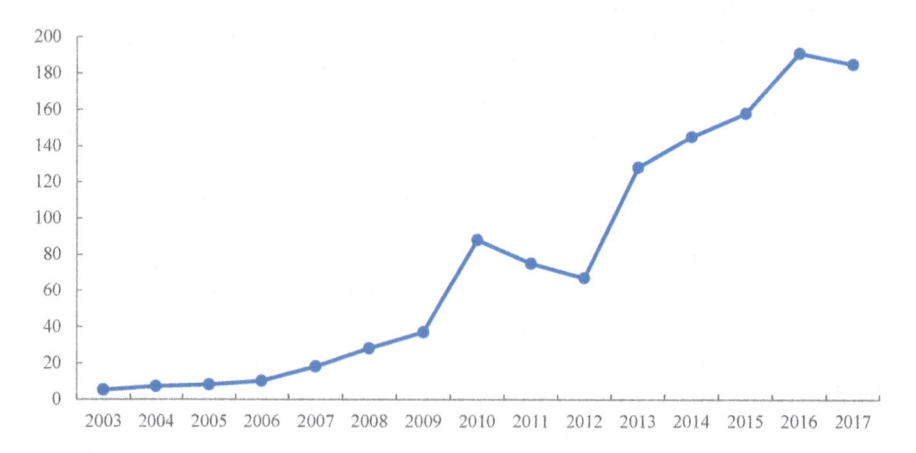

Fig. 1 Quantity of literature of research on cruise industry in China

and Luo (2011) and Yang (2015) summed up and analyzed the development of the cruise tourism, cruise tourists and the influence of cruise tourism, etc. in foreign literature. Shen (2011), through analysis on the research on the cruise industry in China, found that such researches had mainly focused on the basic concepts, the development of the cruise industry both at home and abroad as well as relevant countermeasures, planning and construction of cruise terminals, cruise centers and cruise home ports, types of cruise ships and cruise manufacturing, etc., which paid no attention to the core problems and hot issues in the cruise industry, with inadequate depth and monotonous methods and other defects on the whole. Zhou (2013) found that the literature of research on the cruise industry had mainly focused on the research on the cruise tourism in specific port cities, the training of talents in the cruise industry, and the international cruise industry and the cruise tourism markets, and pointed out that the future development trends of academic articles would include a wider range of research contents, more applications of the method of positive analysis, and the research on newly-formed hot issues such as characteristics of cruise tourist consumers, cruise tourism routes and cruise tourism in specific port cities. In addition to the researches on the overall study of the cruise industry, some scholars also made discussions from certain part of the perspectives. Cai and Shi (2014) separately reviewed the literature concerning the cruise ports in China, and found that the literature had mainly focused on the influence of cruise ports on regional social and economic development, the tourism competitiveness, the planning of cruise ports, the site selection, and the construction of cruise ports and evaluation index systems etc.

3 Research on the Influence of Cruise Industry and the Regional Development

According to the existing literature, many scholars have carried out their research on the cruise industry from a macro perspective, especially with the concerns about the characteristics and influence of the cruise industry, the development of the cruise industry in coastal port cities and other aspects, which has functioned as a good guidance to a complete understanding of the overall development of the cruise industry in China.

3.1 Characteristics and Influence of Cruise Industry

Based on large luxury cruise ships, the cruise industry is a composite industry covering transportation, port service, tour and sightseeing, entertainment and shopping and other industries. Xu and Gao (2010) summed up the elements of the supply chain of the cruise industry, including cruise lines, cruise products, travel agencies and tourist flow systems, organically integrated the core ideas of supply chain into the cruise industry, the new type of industry, according to the characteristics of cruise tourism products, analyzed the composition of the cruise industry and interrelationship between each link from the perspective of supply chain, and put forward corresponding measures for the optimization of the supply chain of the cruise industry.

Along with the rapid development, cruise tourism has exerted ever greater influence and many researchers have been attracted to study the impacts of the cruise industry on economy. In his research on the regional economic contribution driven by the home ports and the ports of call, Luan (2008) determined their respective amount and calculated the ratio between the two. He also forecast the ratio of the most developed American home ports to the ports of call and analyzed the trend by applying neural network theory and building a back propagation (BP)-based neural network. Zhang (2008) applied regression analysis method to analyzing the development data of cruise tourism in North America 2000–2006, and studied the relations between the development of cruise tourism and the regional economic growth, i.e. economic effect of cruise tourism. The results showed that the performance of cruise tourism has impacts on economic growth, which in turn exerts influences on the performance of cruise tourism, i.e. there is a two-way cause and effect relation between the two. Starting from cruise product and characteristics of cruise economy, Sun (2017) studied how international home ports impacted the regional economic development and investigated how regional economy was driven by the home ports based on the growth pole effect, investment multiplier effect, and the regional industrial and spatial restructuring effect of the home ports. The results suggested that international home ports have created new economic growth pole, formed new spatial structure of regional economy, and generated spatial neighborhood effect.

By analyzing research literature, we found that some researchers also showed an interest in the influence of cruise tourism on the environment. Let's take some of the cases as an example. Xie and Li et al. (2010), based on life cycle theory, discussed how the environment could be impacted by such processes as construction and operation of cruise facilities, shipment and distribution, consumption and utilization, as well as waste disposal. They also suggested some basic ideas on environmental pollution control. Du and Li et al. (2016) studied onshore power supply facility at the Cruise Terminal in Hamburg, Germany, and discussed the experience in engineering, legislation, policy making, financing and social participation, which would definitely be inspiring for onshore power projects at cruise terminals of China. Wang and Huang (2015), from the perspective of cycling economy, discussed environmental pollution control in cruise tourism on micro, intermediate and macro scales. However, they just discussed the issues generally, without specific case studies.

3.2 Regional Development of Cruise Industry

As for the development of cruise industry in different regions, researchers performed analyses from various perspectives. Hu and Chen (2004) applied industrial cluster and core competence theories to analyzing feasibility and necessity of development of Shanghai as a terminal, pointed out restrictive factors for developing cruise economy, and proposed a dynamic mechanism to nurture industrial cluster of the cruise industry in Shanghai. Zhang and Fang (2006) used SOWT to analyze conditions of Qingdao terminal to develop cruise tourism, and concluded that Qingdao had proper natural and economic conditions in addition to government support and market demand, but needed better hardware facilities, higher reputation and stronger competitiveness over adjacent ports. They predicted the development trend of Qingdao terminal and believed that cruise tourism would be a new type of tourism product of Qingdao. Li and Zhou et al. (2015) applied SWOT-PEST analysis method to studying the conditions of Nanjing to develop the cruise tourism, discussed response solutions to deal with major restrictions, such as conditions of Nanjing cruise terminal and supporting service facilities, "soft environment" (e.g., policies, culture, and products), as well as regional competition of the terminals, thereby suggested that the development of Nanjing's cruise economy should focus on the characteristics of river port by following a development mode of integrated "city of cruise" featured with diversified operation under the direction of the government, with the development of industrial and port alliance being highlighted at the same time.

Over the past few years, while the coastal cities of China grasped opportunities of this new market and took great efforts to facilitate its development, the growth of cruise industry has also been restricted by various factors. Xie and Zhao et al. (2012) suggested that the growth of cruise tourism was influenced mainly by consumers, tourism resource, and cruise terminal condition, and described positioning and overall pattern of cruise terminal development in Guangdong based on the development environment, i.e. Guangzhou should nurture a layered cruise terminal development pattern that includes international and regional home ports and multiple ports of call. To facilitate better and faster growth of cruise industry, many scholars proposed suggestions and countermeasures. Ye and Sun (2007) applied SOWT method to analyzing the cruise tourism in Shanghai and put forward several suggestions including identifying target market, enhancing advertisement, and improving customs clearance system. Zhang and Cheng (2012) offered their suggestions on how to deal with small market, generally low revenue, incomplete policies, isolation with regional development and other problems of China's cruise tourism, such as establishing a coordinated policy making and planning mechanism, executing industry support policies, forming a systematic "overarching design" featuring mutual coordination and full coverage, as well as actively nurturing actors of the cruise market, in order to strengthen international competitiveness of the cruise industry in China.

3.3 Competitiveness of Cruise Terminals

Numerous researchers have analyzed the competitiveness of ports with different measuring and statistical methods in order to further understand the competition in the cruise industry in China. Yu (2008) analyzed the home port of Dalian and major domestic competitors in terms of terminal, resource, location, and traffic, and comprehensively evaluated the competitiveness of the home port of Dalian by building an AHP model and index system dedicated for analyzing the competition among home ports. He also put forward suggestions on the issues during the development of home port of Dalian.

Zhu (2010) applied factor analysis method and built an index system to evaluate the tourism competitiveness of cruise ports, covering the tourism service performance, the urban average per capita financial strength, the development potential of cruise ports, and the tourism resources of the port cities. He used this index system and quantitatively evaluated 15 port cities of China, and ranked them in this order: Shanghai, Xiamen, Zhuhai, Shenzhen, Tianjin, Qingdao, Guangzhou, Sanya, Ningbo, Dalian, Shantou, Zhanjiang, Yantai, and Haikou and Beihai.

Based on entropy-weight TOPSIS method, Nie and Dong (2010) established an urban cruise tourism competitiveness evaluation system, which includes six Level-1 indexes (resource reserve, market size, economic strength, environmental protection, location, and development potential) and 22 Level-2 indexes, and performed positive analysis of cruise tourism competitiveness of nine port cities. The results showed that Shanghai, Tianjin, and Shenzhen ranked the top 3, followed by Guangzhou, Xiamen and Dalian. Qingdao Ningbo and Haikou presented weaker competitiveness.

Cai and Niu et al. (2010) built an index system to evaluate the development potential of cruise industry, covering six aspects: port support condition, tourism resource potential, tourism market potential, operation capacity of tourism enterprise, support to urban economy, and transportation guarantee. Based on these indexes, they used factor analysis method to perform quantitative comparison and potential of the cruise industry of eight Chinese cities, and classified them into three categories: advantageous zone topped by Shanghai, comparatively advantageous zone represented by Shenzhen and Tianjin, and ordinary advantageous zone represented by Ningbo, Qingdao, Xiamen, Dalian, and Haikou. Generally, the top ranked port cities are basically the same in different studies. Shanghai, Tianjin, and Shenzhen enjoy sound development of cruise tourism and thus possess strong competitiveness in this regard. To improve the market competitiveness, the application of marketing strategies is also important in addition to the product quality and price attraction.

Qiao (2010) analyzed the problems in Shanghai's cruise tourism industry from the standpoints of tourists, travel agencies, and cruise lines, and designed marketing strategies for cruise lines, particularly targeted to Shanghai market, including conducting sufficient market survey, accurately identifying target market, enhancing advertisement for cruise tourism, actively fostering Shanghai market and combining marketing resources, in order to exploit Shanghai market.

4 Research on Cruise Operation and Route Planning

4.1 Cruise Operation and Management

As an emerging industry with booming development momentum, the market of cruise has increasingly attracted more attention of the academic community. Many scholars have, from the perspective of this market, studied on the ticket price, products, and market competitiveness, etc. of cruise ships. Currently, the Chinese cruise market is haunted by vicious competition with low price. Sun and Ye et al. (2016), by means of telephone interview and market review, discussed about the pricing mechanism of the cruise market in China, which uncovered the reasons leading to the price fall, including the dominance of package cruise tourism products, the participation of tourism intermediaries in the competition as a main body, the monopolistic competition market, and the fierce price competition and environment uncertainty. In this paper, the SCP Model is also introduced to evaluate the effectiveness of price competition of the cruise market from three perspectives, i.e., the market structure, the market behavior and the market performance, and the results of the research verified that current price competition had turned out to be excessively underpriced completion, perplexing the domestic cruise market in a "dilemma of low price".

As for the pricing of cruise products, Sun and Feng (2013) put forward a simple two-phase framework for the research of demands on and price adjustment methods for the cruise market. Based on the assumption that the reserve price of cruise passengers were distributed uniformly within the margin of price, the linear demand function for different periods obtained, which lays the foundation for the determining of the number of voyages in the future and ticket price for different periods. This method is beneficial for cruise companies exploring the maximum reserve price of customers and the dynamic change tendency of the market scale. Differently, Liu and Yang et al. (2016) established the price and population forecast related analysis model and the maximum expected ticket revenue model by the comprehensive use of the interpolation method, curve estimation method, correlational analysis method, descriptive statistical analysis, canonical correlation analysis and other techniques. By use of SPSS, MATLAB, LINGO and other software for programing for solutions, they forecast the weekly population of reservation during each voyage and expected ticket revenue of the voyage, analyzed the relationship between the price of reserved seat and the population of reservation and the intended population of reservation, perfected the price list of reserved seats for each voyage and formulated optimal upgrading schemes. He and Zhu et al. (2016), by use of the classical incremental method, obtained the weekly population of intended reservation, established the relationship model between reservation population and ticket price and forecast the weekly average price of reservation of the cruise companies. Besides, by use of the Markov chains forecasting method, they established the tourist aspiration model of upgrading different seats and obtained the increment of expected ticket revenues under different upgrading schemes.

4.2 Cruise Route Planning

In the competition on the cruise market, in addition to the influence factor of price, rich and unique cruise tourism products are significant factors attracting tourists. At present, the research on cruise products by the academic community mainly focuses on onshore routes and onshore sightseeing products. With respect to cruise routes, Sun and Wu et al. (2015) held that China was now at the initial development stage of the cruise industry, when the settings of routes were relatively single. The cruise routes of home ports were mainly located in Northeast Asian area of South Korea (the departure ports of which mainly include Tianjin, Qingdao and Shanghai), Southeast Asian area of Vietnam (the departure port of which is mainly Sanya) and cross-strait area (the departure port of which is mainly Xiamen) and the layout with distinctive characteristics, diversified themes and long-and-short play had not yet come into being. Xu and Gao (2010) pointed out that cruise ship was inherently a kind of tourism product; with compete and considerate star-level guest rooms, shopping, recreation, meeting and other facilities onboard, tourists could thoroughly enjoy onboard living and entertainment facilities, which were the main constituent part during their maritime travel; in midway, the cruise ships would pull off, for the purpose of sightseeing, shopping or visiting and when returning to the departure port, it meat that the maritime travel was about to be over. In terms of the research on onshore cruise products, Duan (2013) put forward that after going ashore, cruise tourists would usually be active within the radius of 2-h driving. Taking Zhejiang as an example, Duan designed the tourism products therein. In Zhejiang, the boutique tourism products appropriate for cruise tourists include sightseeing of the beautiful archipelago Zhoushan Island, visiting of former residences of celebrities in Xikou, Ningbo City—the dominated oriental port, tour of Yandangshan World Geopark in Wenzhou, tour of Jiaxing—an ancient cultural town, and tour of Tiantai Mountain—a famous mountain on the horizon. Ye and Sun (2007) mentioned in their paper that the onshore activities of cruise tourists were mainly composed of sightseeing and shopping; in addition to various recreational activities that were enjoyable on the cruise, if the destinations of cruises on berth could boast beautiful scenery, unique sightseeing experience and diversified shopping options, such cruise products would become much more attractive for Shanghai residents. Sun and Wu et al. (2015), according to the standards of classification based on product properties issued by Royal Caribbean Cruise Lines, made analysis and found that onshore products along cruise routes could basically be classified into nine classes of recreation and sightseeing, adventure tour, incredible food journey, performances and entertainment, exploration of wild fauna and flora, beach and water sports, snorkeling and diving, and flying sightseeing and golf, among which, recreation and sightseeing/urban tourism was the dominant product, together with beach-water sports, performances and entertainment, adventure tour, performances and entertainment, exploration of wild fauna and flora and incredible food journey.

5 Research on Cruise Customers

As the main object of experience of cruise tourism, tourists play an essential role for the development of the cruise industry due to their demands, motivation, behaviors and degree of satisfaction. Among all cruise-related research documents, Ye and Sun (2007), considering the demands and motivation before cruise tours, aimed at the development status of cruise tourism in Shanghai, analyzed the consumption demands of Shanghai cruise tourism and pointed out that Shanghai residents were generating increasingly enhanced ability of tourism consumption and more dense interest on cruise tourism, with the main activities after going abroad being various kinds of sightseeing and shopping. Sun (2015), directing at the status quo of cruise tourism consumption in China, based on the definition and features of cruise tourism, found that the main motivation driving the consumption of cruise tourists was presented as the motivation of leisure consumption, the motivation of holiday consumption and consumption-aimed motivation and put forward the strategies for enhancing the intents of Chinese consumers buying cruise tourism products with pertinence. With respect to the methods for research, in addition to the theoretical analysis based on the status quo of development, some scholars also analyzed tourist demands using mathematical statistics methods. Sun and Zhou (2017) realized the view of consumers on cruise tourism, their understanding channels, consumption behaviors, consumption intents, the reasons why consumers would select cruise tourism and other problems through questionnaires and found that the majority of people under survey would participate in cruise tourism together with their families and friends, they realized the mode of cruise tourism through TV broadcasting and internet, etc. and their main purpose included relaxing on holidays and sightseeing; besides, the strategy of promoting the innovative development of the cruise industry in Guangxi was further put forward under the driving of the "Belt and Road" strategy.

In regard to the research on tourist behaviors in cruise tourism, Liu and Meng (2017), through questionnaire, selected tourists that had and had not experienced cruise tourism as the research objects, analyzed the consumption behaviors in the cruise tourism market in Xiamen and found that monthly household income was the foundation to further dividing the cruise tourism market, market price was the key factor of decisions made about cruise tourism and safety was the essential factor during cruise tourism, on the basis of which, relevant countermeasures for the development of tourism in Xiamen were put forward. Similarly, Xu and Wu (2015) also selected the means of questionnaire, for analyzing the selection of tourism opportunities, tour motivation, information obtaining channels, consumption level and structure and other behavior characteristics of cruise tourists in Shanghai and found that at present, cruise tourism mainly attracted young and middle-aged families with middle or higher income in cities of home ports and surrounding cities, as well as enterprises holding meetings, incentives and business travels; the time of tour was mainly during holidays and festivals and paid vacations except for the Spring Festival; and travel agencies remained to the main broadcasting and sales agents of cruise tourism products. Huang and Qiu (2014) selected Chinese tourists on the "Voyager of the Seas" as respondents and in combination with questionnaire and interview, made research and found that these tourists tended to present the consumption behaviors of weak consuming willingness on rechargeable restaurants, high preference for inside staterooms with larger cost performance ratio, strong intent on duty-free goods, etc. and put forward the suggestions for the improvement of cruise consumption behaviors, such as additionally arranging large-area featured Chinese restaurants, promotion on advanced cabins, rechargeable services, introduction of more high-end brands, etc. For the purpose of having a more comprehensive understanding about consumption behaviors of Chinese and foreign cruise tourists, Qiu and Xia (2017), by means of questionnaire, interview and other methods, studied on 387 Chinese and 318 UK and American tourists. According to this research, the pre-tourism decision-making phase was mainly affected by demand recognition, tourism motivation, acknowledgement of cruise tourism, experience of cruise tourism, relevant matching policies, etc.; the difference of consumption behaviors during cruise tourism was mainly due to the cultural quality of Chinese and foreign cruise tourists and China should pay attention to the cultivation of domestic cruise culture; and during post-tourism evaluation, the industrial maturity of Chinese and foreign cruise industries, the quality of cruise products, aesthetic experience of cruise tourism, marketing methods for cruise products, etc. would make large difference.

In order to confirm the degree of satisfaction after cruise travels, Bao (2014) selected tourists on the Star Pisces of Star Cruises as the objects of study and surveyed and analyzed the degree of satisfaction thereof about various services on the ship. It was found that the overall satisfaction was high, especially of recreation and guest rooms, followed by restaurants, while front desk service was the least satisfactory item. Based on these results, management measures of cruise services to enhance customer satisfaction were put forward, including the reinforcement on the management of customer relationship value, improving facilities and equipment on the Star Pisces and enhancement of the quality of cruise staff. Guan and Wu (2014), through field survey and questionnaire, analyzed the satisfaction of cruise tourists departed from the home port of Shanghai, respectively tested and evaluated the degree of satisfaction of tourists about various facilities and services on the cruise ship and studied on the difference of satisfaction of cruise tourists with different group characteristics. Based on the study results, it was found that different tourist groups basically had uniform opinion as for the degree of satisfaction; specifically, tourists were most satisfied about the services provided by the staff of various departments and the routes of cruise and onshore journey determined by the cruise were accepted; however, as for as-built internet café, libraries and such kinds of associated hardware and facilities on the cruise ship, customers were less satisfied. Meanwhile, the rate of re-visit and recommendation was relatively high and the overall impression of tourists on cruise tourism was good. Kong and Liu (2015), based on the survey on the satisfaction of tourists from home ports in China issued by Royal Caribbean Cruise Lines, in combination with their personal experience on the "Mariner of the Seas" ship, found that the degree of satisfaction of international cruise services was far higher than that of traditional vacation modes and the crew members of international cruises were apt to industriously enhance the quality of services in order to bring an enjoyable and unforgettable journey, at the same time of increasing personal incomes. Sun and Hou (2017) established the satisfaction index system of cruise tourists from home ports by the method of multi-index evaluation and tested and evaluated the degree of satisfaction of cruise tourists departed from Shanghai based on this index system, thus recognizing the gap with expectation of individual index; it was found that in Wusong Port of cruise in Shanghai, indexes with relatively obvious gap from expectations included internet service/WIFI, shopping/duty-free shops, customs clearance/security inspection efficiency, foreign exchange, children areas, going on board, rest rooms, parking lots, diversion of people/vehicles, registration process, post office/toilet/public telephone, going ashore and public transport/passenger transport, and the countermeasures and measures to enhance the degree of satisfaction of the home port of Shanghai were mentioned. Sun and Ni (2017), standing on the perception of cruise tourists, on the basis of document analysis, expert review, guided survey, questionnaire and other methods, constructed the index system of satisfaction of onboard services and facilities and equipment; besides, with international cruise ships departed from the home port of Shanghai as the object of study, they found that there was a large gap between customer satisfaction and expectation as for internet service, laundry service, onboard shopping, education/learning/training courses, food (Chinese food), fitness clubs, libraries, beauty salon, business/conference center, game/competition, bar and passenger communication/interaction and other indexes, of which, attention should be given.

6 Other Researches

6.1 Cruise Manufacturing

In addition to the above research perspectives, there are also scholars have paid attention to other perspectives in the cruise field, such as cruise culture, cruise manufacturing, laws and regulations on cruise ship and talent training, among which only a few of literature was involved. In terms of cruise manufacturing, Xu and Sun et al. (2008) conducted careful research on numerical sediment model test of the project of Tianjin Port International Cruise Terminal. Wang and Zhang (2008) discussed about the planning and design of cruise home port, taking Hong Kong Kai Tak Cruise Terminal, Los Angeles World Cruise Center, Barcelona Cruise Terminal D and Vancouver Cruise Terminal as examples. Sun and Wang (2009) analyzed the current issues and development trend of the world cruise manufacturing industry and the necessity and feasibility of developing cruise manufacturing industry in China, and put forward several proposals in this regard.

6.2 Cruise Security

In terms of cruise security, Luan (2016) analyzed the types and characteristics of emergencies during cruise transportation, and further discussed the building issues of emergency system of cruise ship in terms of coping approach and legal undertaking. Wang and Zhou et al. (2012), from the perspective of emergency evacuation at cruise terminal, built cellular automaton model for escaping, which simulated the situation of personnel evacuating in emergency circumstances, and put forward optimizing strategies in the number and width of evacuation exit, evacuation routes and evacuation management. In fact, the social impact of cruise tourism should be also reflected in the cultures, heritages, communities, residents, employees, passengers and even other tourists of the destination. However, at present, China has seen no related research results.

6.3 Talent Training

In terms of cruise talent supporting, Zhao (2009) studied the current existing problems in the education of cruise tourism talent in China, and expounded the characteristics, target location and principles of course design of cruise tourism talent training in institutions of higher education, and then presented the training mode for cruise tourism talent in institutions of higher education in China. Ge (2010), in combination with the reality in China, put forward talent training mode that to be dominated by government, promoted by the industry, engaged by the industry, institutions of higher education and training and internationally compatible.

6.4 Laws and Regulations and Miscellaneous

In terms of laws and regulations, Shao and Zhang (2007), after analyzing the factors restricting cruise inspecting efficiency, pointed out that relatively independent cruise control regulations should be formulated for cruise inspecting, visa policies for cruise passengers should be properly loosened and convenient visa procedure should be carried out for mariners and tourists who come to China by cruise ship. Ma (2008), from the legal point of view, analyzed many related problems such as passenger transportation by sea, cruise building, planning of cruise home port and customs clearance for cruise passengers involved in the cultivation and development of cruise economy in the Bohai coastal region and put forward corresponding countermeasures. In the end, in terms of cruise culture cultivation, Shen and Min (2007), with traditional connotation and new characteristics of cruise culture as the commencement, analyzed the profound implied meaning and new characteristics expression of the culture, and offered advices on developing the cruise industry and culture in our country.

7 Conclusions and Prospects

After analyzing the important results of the research on cruise industry in China, it is found that in recent years, with the rapid rise of the cruise industry in China, cruise tourism research begins to enter a relatively prosperous period and the research contents become more specific and detailed, involving demand forecasting of cruise tourism, allocation and pricing of cruise cabin seats, planning of cruise route, seasonal characteristics of cruise market, characteristics of on-line attention to cruise tourism, competitiveness evaluation of cruise terminal, satisfaction evaluation of cruise home port, laws and regulations on cruise ship, training of cruise talents, and environmental protection and security of cruise ship. According to the present development status and domestic academic research situations of cruise industry, the research directions in the future are suggested as follows:

(1) The residents' attitude to the development of cruise industry: During the process of rapid growth of cruise tourism, the attitude of residents of port cities to cruise tourism plays an important part in the development of cruise economy. Targeted cruise products meeting travel demand can be designed, the potential of cruise market can be further tapped and the development of cruise industry can be better facilitated only when the perception and demand intention of our country's residents toward cruise tourism are fully understood.

(2) The expanding and marketing pattern of cruise market: Buy-out of part of cabin seats and chartering are marketing patterns of cruise market in China and 90% of cruise products are distributed through chartering, and such marketing pattern will lead to the sale of cabin seats at reduced price and thus result in vicious competition of low price market. Therefore, in the future, the academic research may pay attention that whether the marketing pattern of cruise industry in China can be innovated to get rid of existing low price predicament.

(3) The environmental protection and security issues of cruise ship: Cruise security is a very noteworthy issue. During voyage, cruise ship might be affected by the factor of force majeure and thus lead to accident. The consequences would be disastrous once cruise ship suffer shipwreck. Some scholars have noticed the security issue of cruise ship and proposed that related personnel can guide the tourists to purchase cruise travel insurance. The academic research may pay attention to cruise security issue in the future to further guarantee the travelling of tourists.

(4) The detailing of research methods and interdisciplinarity: Researchers can mostly adopt diversified research methods during cruise research, such as conducting case analysis and empirical research through combination of qualitative and quantitative method and by using the method of mathematical

statistics and data analysis, so that the cruise industry can be researched in a more scientific and comprehensive way. In addition, cruise industry is a composite industry involving many industries, and the industry chain will involve production and manufacturing, product design, promotion and marketing, etc., so interdisciplinarity must be emphasized during the research of cruise industry, such as comprehensive application of geography, marketing, psychology, ecology, environmentology, sociology, the science of law, and only in this way the realistic problem of the industry development can be recognized and resolved in a more comprehensive way.

Chapter 7
The Economic Contributions of Evaluation Indicator System to Shanghai's Cruise Industry

Huang Huang

1 Literature Review

In recent years, the cruise industry in Shanghai has developed very well. With the rapid development of cruise industry, its role in the economy growth in Shanghai has been increasingly remarkable. However, the current studies on the economic effects of the cruise industry on Shanghai are mainly based on qualitative analysis; there are few studies on the economic contribution of the cruise industry to Shanghai. This phenomenon is so unfavorable to the insight into the important factors affecting the contribution to local economy that the healthy and rapid cruise economy development of Shanghai is limited. This study explores and analyzes the economic contribution of cruise industry, and establishes an evaluation indicator system.

1.1 Economic Contribution of Cruise Industry from the Theoretical Perspective

Cruise economy is a special tourism economy form, which has both the outstanding features of traditional tourism economy and multiple distinct characteristics. Therefore, both the generality and specificity of cruise economy shall be taken into consideration during the study on the economic contribution of cruise industry. Adjustment and improvement shall be made by taking full use of recent theories for

---

Research Directions: Cruise Tourism, Residents Leisure, Pension Travel.

H. Huang (✉)
China Tourism Academy, Beijing, China
e-mail: 2617533@qq.com

the evaluation on the influential effect of tourism economy according to the structure of oligopolistic enterprises, "flag of convenience"-oriented ship registration structure, competition and cooperation relationship between cruise and city, global ship supply procurement network, obvious seasonal fluctuations and other characteristics of cruise industry. A theoretical system for evaluation on the economic contribution of cruise industry will be established.

1.1.1 Characteristics of Cruise Economy

The divisions of industry in various countries in the world are based on the similarity of production activities. As indicated in the current standard of China, the *Industrial Classification for National Economic Activities* (GB/T 4754-2011), "National economic activities shall be divided based on the principle of homogeneity, that is, each industry classification is based on the nature of the same economic activity, not on the establishment, accounting system or divisional management", all national economic activities are divided into category, large class, medium class and subclass.

Nevertheless, tourism is divided as the complementarity of tourists' consumption. Tourists inevitably generate such consuming activities as "board and lodging, travel, tour, shopping and entertainment" in the process of traditional tourism and such activities as business, regimen, leisure, health care and study in new tourism products and new forms. Such tourism activities are classified into accommodation and catering industry, wholesale and retail sales, resident service industry, culture and art industry, entertainment industry and other industries in the *Industrial Classification for National Economic Activities*. However, such industries also provide local residents with products and services, thus it is difficult to effectively distinguish the economic activities serving tourists and residents. Therefore, the value added of tourism is dispersed over various industries. Except for travel agency industry, it is difficult to know the development of tourism directly from data and more difficult to directly calculate the economic contribution of tourism.

Cruise industry is a special tourism covering various products and forms in tourism. The consumption of cruise tourists, crew and enterprises is extremely complex and multivariate and is scattered in various industry categories of national economy. Therefore, it is also difficult to learn about the development of cruise industry directly from data. The evaluation on the economic contribution of cruise industry may be completed by means of special data source, theoretical method and indicator system.

Compared with ordinary tourism, cruise industry has some outstanding features, which should be taken into account when evaluating the economic contribution of cruise industry. A cruise may be functionally regarded as a large tourist resort, where tourists may spend a vocation on the cruise and enjoy such leisure and entertainment services as catering, accommodation, entertainment, sports, health, shopping and culture. However, seen from economic contribution, cruise and tourist resort are quite different. The tourist resorts in Shanghai, regardless of the ownership structure, are local enterprises registered in Shanghai, therefore all economic activities in such tourist resorts are included into the local value added in Shanghai and all income thereof into the gross revenue of tourism industry in Shanghai. Most of cruises, however, hang flag of convenience and are registered abroad, therefore the owners of cruises are several large multinational companies. Therefore, Chinese tourists' participation in cruise tourism is outbound tourism and relevant consumption belongs to consumption abroad. Even though tourists pay in RMB, the economic activities on cruise will not be included into the local value added in Shanghai; and the income of cruise enterprises will not be counted as the gross revenue of tourism industry in Shanghai. Therefore, the method to evaluate the economic contribution of traditional tourism cannot be applied to the evaluation on that of cruise industry. The economic contribution of cruise industry to Shanghai is mainly reflected in such aspects as the consumption of tourists and crew in Shanghai, the commodities and services procured by cruise enterprises in Shanghai, the business office expenditures of cruise enterprises in Shanghai, the cruise-relevant infrastructure and public services provided by the government and relevant fixed assets investments.

It can be concluded that cruise industry cannot make economic contribution to Shanghai unconditionally due to its specialty, that's why the cruise tourism is prosperous but local economy develops slowly in many countries. It is necessary for Shanghai to take multiple measures to increase the economic contribution of cruise industry to Shanghai on the premise of the scientific evaluation on the economic contribution of cruise industry.

1.1.2 Temporal and Spatial Boundaries of Economic Contribution of Cruise Industry

(1) Temporal boundaries of economic contribution of cruise industry

Generally, the cruise industry with Shanghai as the home port has definite temporal boundaries from the time when tourists complete exit formalities and board the cruise to the time when they successfully go through entry formalities and come back to Shanghai. However, such definition has ignored the consumption of tourists travelling to and from Shanghai and their consumption in Shanghai before and after cruise tourism, thus it may underestimate the economic contribution of cruise industry to Shanghai. The practice internationally widely applied is to classify all supplementary economic activities in cruise tourism into cruise industry for such activities are not likely to happen if there is no cruise tourism. Therefore, if the main trip purpose of tourists is cruise tourism, all activities happened in the process of this travel belong to cruise industry.

(2) Spatial boundaries of economic contribution of cruise industry

Since cruise industry in a comprehensive industry and many industries have global value chain, the consumption of cruise tourists and cruise enterprises cannot always reflect the actual economic contribution of cruise industry to Shanghai. Taking the fuel oil of cruise as an example, there is no oil exploitation industry in Shanghai, thus most of the amount paid by cruise enterprises to Shanghai fuel oil enterprises is used for oil import and a little is for the consumption in Shanghai. Similarly, if a cruise tourist consumes a bottle of imported grape wine, most of the consumption is transferred to relevant foreign country, forming a "leakage loss" of final consumption. Therefore, the direct use of the consumption data from cruise enterprises or tourists may overestimate the economic contribution of cruise industry to Shanghai.

Seen from the practice of the evaluation on the economic contribution of cruise industry in various countries in the world, the "leakage loss" of relevant imported products shall generally be deducted from the consumption data to obtain the actual local output value and calculate such indicators as value added, employment, taxes and multiplier effect. Theoretically, the proportion of the "imported" products among "intermediate products" and "end-use products" may be estimated from the input-output table of Shanghai; and the output value in Shanghai may be obtained by multiplying this proportion by the direct consumption data of cruise tourism. However, since the divisions of industry in the input-output table are not detailed (only 42 industries) and the products consumed in cruise tourism, a special industry, may not be the "average commodities" consumed by all urban residents and cannot replaced by the mean value, survey and other methods may be required to obtain the "leakage loss rate" of direct consumption during the study on the economic contribution of cruise industry. Generally speaking, most of the income from such port services received by cruise enterprises as tug, navigation, berthing, loading and unloading, and such services enjoyed by tourists ashore as catering, accommodation, urban transport and entertainment belongs to the revenue of Shanghai. The economic contribution of the commodity consumed by cruise enterprises and tourists depends on whether the commodity is produced in Shanghai; while the economic contribution of such trans-regional transportation as HSR and aviation depends on the ownership structure and profit distributing mechanism of railway companies, airlines and other enterprises.

Since the consumption in cruise industry may come from the whole country or even the whole world, the evaluation on the economic contribution of cruise industry by different spatial scales may draw different conclusions. In general, regions with larger spatial scale may retain more economic effect of cruise. For regions with equal scale, economic entities with diverse economic structure and strong industrial competitiveness may have more profound economic effect of cruise. As one of the most economically developed cities in China, Shanghai has great potentiality in developing cruise economy. This study focuses on the evaluation on the economic contribution of cruise industry to Shanghai. When the data for such evaluation gradually get complete, relevant models may be used for the evaluation on the economic contribution of the cruise industry to Shanghai, the whole country and the whole world.

1.1.3 Production Substitution Effect of Cruise Economy

According to economic theoretical analysis, the increase of demands for cruise tourism in Shanghai may speed up cruise industry development, but may not bring about the growth of the economic aggregate in Shanghai. The production factors, such as land, capital and labor, for cruise industry development, may come from the recycle of idle production factors or the introduction of new production factors, which may contribute to the growth of economic aggregate; such production factors may also from the occupation of other industrial factors, which may lead to the recession of other industries. Taking hotels as example, if the tourist accommodation reception capacity in Shanghai develops slowly and cruise tourism coincides with the rush season in this city, the development of cruise tourism in Shanghai may have occupied the existing tourist accommodation reception capacity, raised the prices of hotels and constrained the development of urban tourism, which may limit the promotion of the growth of the economic aggregate of Shanghai. On the contrary, if the tourist accommodation reception capacity in Shanghai can adapt to the growth rate of cruise tourism, cruise industry development may increase the economic aggregate of Shanghai on the premise of stable tourist price.

Seen from the actual conditions of Shanghai, the cruise industry is now at initial development stage; the ratio of cruise tourists (607,500 person-times in 2014, averagely 1664 person-times a day) to the number of beds (96,000 beds in 2013) in star-rated hotels in Shanghai is not high; and a considerable proportion of tourists are from Jiangsu, Zhejiang and Shanghai, thus they do not live in such hotels. Therefore, cruise industry has small pressure on the tourism reception facilities in Shanghai; and the production substitution effect of cruise economy did not appear. Nevertheless, the rush season and off-season of cruise tourism fluctuate greatly, therefore the arrival at and departure from the port of cruise may cause short-term impact on the city. With the achievement of the goal of cruise tourists 2 million person-times ahead of schedule, the proportion of tourists from other provinces further increased. There objectively are risks where cruise industry may replace the production factors in other industries. Relevant departments shall effectively respond to the seasonal fluctuation of cruise tourism, increase the reception capacity of cruise tourism and improve the investment climate for cruise industry.

1.1.4 Multiplier Effects of Cruise Economy

Under the economic effect of cruise industry, the consumption by cruise tourists, crew and enterprises of the commodities and service produced in Shanghai can directly speed up the economic growth in relevant industries in Shanghai, and has direct economic effect of cruise industry. Such industries driven by direct economic effect will further promote the growth of relevant industries that provide intermediate inputs for such industries, generating indirect economic effect. The industries under indirect economic effect contribute to the growth of relevant industries. The sum of the above economic effects forms the total economic effect of cruise industry. Each economic effect may calculate such data as the output, employment, taxes and the value added driven, the sum of which is the total output, employment, taxes, the value added and other indicators brought about by cruise economy in Shanghai.

For example, if a cruise tourist has spent CNY 1000 in a hotel in Shanghai, it can be seen from the *Input-output Table of Shanghai (2007)* that 92.8% of the products in accommodation and catering industry, "intermediate products" and "end-use products" are produced locally and 7.2% imported. Therefore, among the commodity and service of CNY 1000 spent by the tourist, the commodity and service of CNY 928 are produced in Shanghai. The direct economic effect of his/her stay in such a hotel is CNY 928. It can be further concluded from the direct consumption coefficient in the input-output table that among the CNY 928, CNY 681 is from the intermediate inputs purchased by accommodation industry from other industries and CNY 247 from the value added of accommodation industry itself. Among the value added, CNY 88 is the payment for hotel service staff, CNY 26 the tax collected by the government, CNY 12 for the depreciation of fixed assets and CNY 120 the operating surplus of the hotel. The products of CNY 681 purchased from other industries may cause several rounds of multiplier effect. It may be concluded, by calculating the complete consumption coefficient in the input-output table, that the final consumption of CNY 928 in accommodation industry can drive several industries to produce the products valued CNY 2699, and generate CNY 342 total remuneration, CNY 163 total taxes, CNY 97 depreciation of fixed assets and CNY 325 total operating surplus. The total value added CNY 928 equals to the final consumption in accommodation industry.

The above mechanism is the multiplier effect in the process of growth of cruise industry. The total economic effect of cruise may be much more than the direct economic effect, which represents the final promoting function of cruise industry on the economic aggregate. According to the practice of the evaluation on the economic contribution of cruise industry in various countries, the calculation of the total economic effect by input-output model has been a standard practice. In addition, Keynesian multiplier model, CGE model and other models may also be applied to the evaluation on multiplier effect.

1.2 Data Acquisition Method for Evaluation on Economic Contribution of Cruise Industry

In order to evaluate the economic contribution of cruise industry in any city, the data on three most basic kinds of cruise industry consumption shall be collected, i.e. the consumption of cruise tourists, crew and enterprises. In addition, additional evaluation data shall be collected and an indicator system established according to cruise industry development in each city.

1.2.1 Economic and Social Relevant Data

First, the evaluation on the economic contribution of cruise industry requires such statistical data as economic growth, industrial structure, employment structure, income distribution, labor productivity, tax rate, rate of inflation, input-output table, inter-provincial trade and foreign trade in Shanghai, which can be obtained from such relevant departments as statistical bureau, tax bureau, business committee, customs and port office.

1.2.2 Consumption of Cruise Tourists

The consumption of cruise tourists is greatly influenced by income, age, itinerary, tourist route, product and other factors, but such influence is not reflected in relevant government statistics. Therefore, the data on such consumption is the most important to the evaluation on the economic contribution of cruise industry and it is the most difficult to obtain. Generally, such data are obtained by questionnaire survey. Tourists are divided into three types: tourists who stay at home port, who do not stay at home port and who call at a port to survey the origin of trip, demographic characteristics, interregional transport, urban residence time, shore tourism projects, tourism consumption structure, tourist satisfaction, revisit intention and other aspects. For tourists traveling with an agency, travel agency may assist in providing partial data.

1.2.3 Consumption of Cruise Crew

The data on the consumption of cruise crew is also collected by questionnaire survey to know their nationality, residence, demographic characteristics, income, urban residence time, shore tourism projects, tourism consumption structure and other information. Sometimes, the cross-regional transportation and urban accommodation and other consumption of cruise crew during transfer at a port can be reimbursed by cruise enterprises. The part of income may be obtained from cruise enterprises or estimated by transportation distance and consumption structure, etc. Since such consumption is relatively centralized and the individual difference is not as significant as that among cruise tourists, the sample size for survey may be properly reduced to improve the efficiency of questionnaire survey.

During such survey, the consumption characteristics of foreign crew, Chinese crew and Shanghai crew shall be distinguished. Generally speaking, the consumption of foreign crew in Shanghai mainly is tourist consumption concentrated on such as aspects as sightseeing, catering, shopping and entertainment. Chinese crew may reside in Shanghai for a long time, thus the living consumption in renting and purchasing durable consumer goods and other aspects is higher. In addition to various consumptions, Shanghai crew may have fixed assets investment such as real estate investment. Though all of such personnel are employees of cruise enterprises, their contribution to the economic growth in Shanghai differs greatly.

1.2.4 Operating Expenses of Cruise Enterprises

As for the operating expenses of cruise enterprises, list of cruise sailings, routes, passenger flow volume, number of crew, port services, fuel, ship supply, repair and maintenance and other data shall be surveyed. All of such data have detailed financial record and may be directly obtained from cruise enterprises. Since the production inputs for cruise enterprises are generally from the whole world, only the part procured in Shanghai may be included during the survey. If cruise enterprises refuse to provide relevant data for the sake of trade secrets and other reasons, such data may be obtained from the cruise ports, cruise industry association, shipping agent enterprises, ship supply enterprises and other servers. In the process of specific survey, it should be noted that cruise enterprises often change suppliers and maintenance enterprises due to frequent transfers at ports all over the world. Therefore, the annual fluctuations in ship supply, repair and maintenance and other consumption are quite significant. The overall development trend can only be obtained by integrating the data for years.

1.2.5 Management Expenses of Cruise Enterprises

Generally, cruise enterprises will set up regional headquarters or branch at large cruise home port and engage professional management and office staff to be responsible for the strategic decision-making, investment and financing, operating management, research and development, financial management, staff training and other aspects. The remuneration of office staff, enterprise office expenses, marketing cost, service outsourcing cost and other costs borne by cruise enterprises may be regarded as their management expenses ashore and directly contribute to urban economy. The above data may be obtained by survey on cruise enterprises or check of the operating data of such enterprises via open sources. For such cruise enterprise headquarters bases as Miami, management expenses have been an important part of the economic contribution of cruise industry. Carnival, Royal Caribbean and other cruise enterprises are considering moving regional headquarters to Shanghai. The role of the management expenses of cruise enterprises in speeding up the economic development of Shanghai cruise will be enhanced.

1.2.6 Cruise-Related Fixed Assets Investment

Except for the huge consumption of cruise enterprises, fixed assets investment can also bring about huge economic effect. The cost of cruise is several hundred millions or even more than a billion dollars, and boasts the most important fixed assets investment in cruise industry. Shipyards may be the important economic pillar in the area where the shipyards are located and promote regional economic development. Huge costs and expenses such as cruise manufacture and decoration may be easily obtained via open sources or from Clarksons and other shipping information database. Attention shall be paid to the difference between the production price and the market price during data analysis. The market price, at which cruise enterprises purchase the cruise, is the sum of the production price and such intermediate costs as freight, taxes and dealer margins. Shipyards can only get the production price; while intermediate costs are charged by other enterprises. Therefore, the market price shall be converted to the production price when evaluating the economic contribution of cruise manufacture to the area where shipyards are located. Since almost all mainstream large cruises in the world are manufactured in Europe and the cities where shipyards are located are always not famous cruise ports, shipbuilding industry is only included as an important part for evaluation on the economic contribution of cruise industry to the whole Europe.

The office buildings and other fixed assets invested by cruise industry can also promote the economic growth in the cities where such assets are located in the form of investment. Such data may be obtained from cruise enterprises, planning bureau, urban and rural construction committee, construction enterprises and other aspects. First, since the investment into large construction projects always lasts for years and the annual fluctuation is significant, the data on fixed assets investment for years shall be collected to calculate annual average investment. Second, only the actual construction cost of projects promotes the economic growth in the cities where such projects are located. The market assessment value or transfer income includes the investment premium of cruise enterprises and does not actually promote the development of other industries, thus cannot be used as the basis for calculating the economic contribution of cruise industry.

1.2.7 Cruise-Related Infrastructures and Public Service

The tourism products and services consumed by cruise tourists in Shanghai can be divided into two categories: catering, accommodation, shopping and other industries mainly provided by tourist enterprises in the market; cruise tourists directly pay relevant amounts to cruise enterprises, forming direct cruise tourism consumption; the other category is represented by museums, art galleries, tourist toilets, municipal roads and public scenic spots, which are mainly provided by governments as public services; even though the provision of such public services causes a lot of government expenditures, cruise tourists do not have to pay for this, forming no direct cruise tourism consumption.

With the continuously increase of cruise tourists in Shanghai, their demands for the tourism infrastructure and public services provided by the government has increased continuously. This will inevitably promote the increase in the cruise tourism-relevant investment and consumption of the government. Similar to the investment and consumption of cruise enterprises, such investment and consumption can also bring direct economic effect and cause total economic effect via multiplier effect. If there is no cruise tourism, such relevant expenditures of the government in public services will not occur. Therefore, such infrastructure investment and public service expenditures of the government shall be included into the evaluation indicator system for the economic contribution of cruise industry. The data on the expenditures of the government in the infrastructure and public services relevant to cruise tourism may be obtained from such government sectors as development and reform commission, construction committee, traffic committee, finance bureau and tourism bureau. It should be noted that there may be great annual fluctuation in the infrastructure investment of the government, therefore the growth trend should be analyzed by collecting the data for years.

1.3 Methods for Evaluating Direct Economic Contribution of Cruise Industry

Seen from the whole tourism economy, two major indicators can generally reflect the importance of tourism in national economy, i.e. the direct economic contribution of tourism. The two indicators are tourist income and the value added of tourism industry (VATI). The data on the tourist income of all provinces and cities in China are published, but only a few regions including Shanghai publish the VATI at the same time. For example, the *Statistical Bulletin for National Economic and Social Development in Shanghai* shows that the tourist income of Shanghai in 2014 was CNY 330.058 billion and the VATI CNY 144.933 billion. Tourist income is a concept of output value and reflects the direction promotion of local economy by tourist consumption when the imported value is deducted. The value of tourism products is from both tourism and the intermediate input industries thereof. In the above example of accommodation, CNY 928 directly consumed by the tourist is tourist income. The VATI is a concept of value added, which is from tourism only and reflects the scale of tourism. The value added in accommodation industry of CNY 247 in the above example is the VATI. It can be seen from the above analysis that tourist income must be greater than the VATI.

Specifically, for cruise economy, there are indicators similar to tourist income and VATI for evaluation on the direct economic contribution of cruise industry. As mentioned above, since the headquarters of cruise enterprises and cruises are basically registered abroad, cruise tourism is outbound tourism, and it is not scientific to directly use the income of cruise enterprises to evaluate the direct economic contribution of cruise industry, the indicator, the income of cruise enterprises, and is replaced by another indicator, cruise tourism consumption. The evaluation the direct economic contribution of cruise industry covers two major indicators, the direct cruise tourism consumption and the value added of cruise industry. The report on the evaluation on the economic contribution released by the Cruise Lines International Association (CLIA) emphasizes the importance of cruise industry (CLIA 2014), and draws a conclusion that the direct cruise tourism consumption was mainly listed and the value added of cruise industry was not. The value added of cruise industry was calculated in the independent studies released by Scholar Braun et al. (2002) and Worley et al. (2013). The data source, content, calculation method and indicator system of the two indicators are introduced below:

1.3.1 Direct Cruise Tourism Consumption

Seen from the evaluation on the direct economic contribution of cruise industry in various countries and regions, direct cruise tourism consumption mainly includes consumption of cruise tourists, consumption of cruise crew and consumption of cruise enterprises.

(1) Consumption of cruise tourists

According to the features of cruise tourists, they can be divided into "home port tourists" and "port of call tourists". "Home port tourists" can be further divided into "tourists who stay at home port" and "tourists who do not stay at home port" according to the length of stay at the port city.

Table 1 shows the total amount and structure of the consumption of global cruise tourists and cruise crew in 2013. The consumption of cruise tourists mainly covers the long-distance transportation to port city, urban transport, accommodation, catering, shopping, urban sightseeing, entertainment, communication and other costs. "Port of call tourists" stay at the port city for a one-day tour. No accommodation and long-distance transportation costs will be incurred. "Tourists who do not stay at home port" mainly regard the port city as a transit hub and only stay in

Table 1 Consumption of global cruise tourists and cruise crew in 2013

Category	Unit	Home port tourists	Port of call tourists	Cruise crew
Accommodation	USD 100 million	10.14	–	–
Long-distance transportation to home port	USD 100 million	38.41	–	–
Catering	USD 100 million	5.41	7.58	3.01
Local sightseeing	USD 100 million	4.77	27.79	1.89
Shopping and others	USD 100 million	8.67	30.69	7.43
Total	USD 100 million	67.40	66.06	12.33
Average consumption each time	USD	316.28	92.00	56.69

Data source CLIA (2014)

the city for only several hours, thus have little consumption of tourism products and services and incur no accommodation costs. By contrast, "tourists who stay at home port" stay in the port city for the longest time often for urban tourism before and after cruise tourism. Therefore, the structure of their consumption is the most diverse and the consumption is the highest.

(2) Consumption of cruise crew

The structure of the consumption of cruise crew is similar to that of cruise tourists and also covers long-distance transportation, urban transport, catering, shopping, urban sightseeing, entertainment, communication and other costs. The most significant distinction is that cruise crews generally do not incur any accommodation costs in port city, nor long-distance transportation. The ratio of entertainment of cruise crew is lower than that of cruise tourists, but the ratio of shopping is higher. If cruise crews are the residents of the port city, their consumption structure is similar to that of local residents. The ratio of local consumption to all of their income is the highest. Furthermore, if cruise crews transfer at a port due to business reasons, long-distance transportation and other costs will be reimbursed by cruise enterprises.

(3) Consumption of cruise enterprises

The consumption of cruise enterprises can be divided into operating expenses and management expenses. Operating expenses mainly include the expenditures for operating the cruise and providing various services for tourists on board; while management expenses mainly covers the business expenditures and office expenses ashore borne by cruise enterprises.

Table 2 shows the specific items of the operation and management expenses of cruise enterprises. For port cities where cruise enterprises have not set up an office, the consumption of cruise enterprises is mainly operating expenses. For home port cities where cruise enterprise headquarters gather, such as Miami, the management expenses have taken up a great proportion of the consumption of cruise enterprises.

Table 3 shows the cost structure of Canadian cruise enterprises in 2012 and reflects the proportion of various consumption expenditure items in the total expenditures of cruise enterprises. Since cruise enterprises has global product supply chain, only the products and services procured by cruise enterprises in the region will be included when evaluating the economic contribution of cruise industry in the region. For the employees of cruise enterprises, cruise crews and employees ashore are generally treated differently. The income of nonlocal cruise crews will not be included in the direct economic effect of cruise tourism. The calculation of their consumption ashore is as stated above. Since the local cruise crews and employees ashore of cruise enterprises live in the port city for long time, their income may be included in the direct economic effect of cruise tourism.

Table 2 Classification of major operating and management expenses of cruise enterprises

Operation expenses	Management expenses
Food and beverage	Advertising and marketing
Marine fuel	Accounting service
Port charge	Legal service
Onboard entertainment	Information consulting service
Ship maintenance and repair	Other professional services
Air tickets for passengers (package tour)	Communication
Ship insurance	Travel and entertainment
Salary of crew	Rent
Other operating expenses	Utility bills
Travel insurance for passengers	Office expenses and remuneration on shore

Data source CLIA—North West and Canada (2013)

Table 3 Cost structure of Canadian cruise enterprises in 2012

Category	Proportion (%)
Business and computer services	20
Fuel	13
Machinery equipment	13
Travel agent	12
Food and beverage	8
Air tickets	7
Port charge	5
Salary	5
Advertising and marketing	5
Ship maintenance and repair	3
Hotel services	2
Other costs	7

Data source CLIA—North West and Canada (2013)

(4) Method for calculation of direct cruise tourism consumption

When the data on the consumption of cruise tourists, crew and enterprises are obtained by all means, the direct consumption of cruise tourism can be calculated by further processing such data. First, even the commodities and services locally procured by tourists or enterprises may be imported, thus can drive local economy to a less degree. Therefore, the proportion of imported value should be calculated by surveying or checking the input-output table and other means to deduct such value from the sum of consumption. Besides, the consumption data obtained by surveying on tourists and cruise enterprises reflects the market price. Compared with the production price, the market price also includes dealer margins and circulation costs. When calculating the promotion of intermediate input industries by cruise tourism consumption, the market price should be reduced to the production price by calculating. Finally, the consumption data obtained by survey are divided by consumption categories. When calculating the promotion of the economy by cruise tourism consumption, the consumption data should be classified into corresponding industries of national economy by the nature of products to get the output of various industries of national economy direct driven by cruise tourism.

1.3.2 Value Added of Cruise Industry

The above-mentioned direct cruise tourism consumption can reflect the direct promotion of economy by cruise industry, and is an important indicator for evaluation on the direct economic contribution of cruise industry. However, since the direct cruise tourism consumption includes the value added of cruise industry and the intermediate inputs consumed by cruise industry, consumption neither reflects the scale of cruise industry, nor can be directly used to calculate the scale of GDP or directly compared with other industries. For example, the accommodation consumption of CNY 928 in the above accommodation case reveals the direct economic contribution of such accommodation to Shanghai, but CNY 928 is not the value added of accommodation industry. Actually, the accommodation only bring about a value added of CNY 247, which can be further divided into laborers' remuneration, net production tax, depreciation of fixed assets and operating surplus.

Therefore, in order to calculate the scale of cruise industry and the proportion of cruise industry in GDP, and compare cruise industry with other industries, the value added of various industries directly contributed by cruise tourism on the basis of the data on the direct cruise consumption should be divided into various industries. The sum of the value added of various industries is the value added of cruise industry. Similar to the concept of Tourism Satellite Account, cruise industry is divided according to the complementarity of cruise tourism consumption, therefore it is difficult to find cruise industry in the *Industrial classification for national economic activities*; it is actually included in such industries as transportation, retail, accommodation and catering. The above-mentioned method to calculate the value added of cruise industry can calculate the relative and absolute scales of cruise industry in various industries. The total of such scales is the scale of cruise industry and the basis for evaluating the importance of cruise industry in national economy.

Besides, the laborers' remuneration, net production tax, depreciation of fixed assets and operating surplus directly brought to various industries by cruise tourism can be calculated on the basis of the laborers' remuneration coefficient, net production tax coefficient, coefficient of depreciation of fixed assets and coefficient of operating surplus of each industry. The total of such indicators is the laborers' remuneration, net production tax, depreciation of fixed assets and operating surplus of cruise industry. In other words, the amount of labor remuneration, tax revenue and operating surplus that can be directly brought by cruise industry to Shanghai can be calculated by dividing the value added of cruise industry into laborers' remuneration, net production tax, depreciation of fixed assets and operating surplus.

Moreover, the quantity of employment in cruise industry can be calculated according to the labor productivity in various industries.

When calculating such quantity of employment, it shall be noted that the calculated quantity of employment can only be the Full Time Equivalent rather than the actual quantity of employment for cruise industry is a highly seasonal industry, labors may work overtime in rush season and be employed in other industries in off-season.

1.4 Method for Evaluating the Economic Contribution of Cruise Industry

The above indicators, such as the direct cruise tourism consumption and the value added of cruise industry can only reflect the direct economic contribution of cruise industry. Since cruise industry can drive the production in various industries, the direct output and resultant output can be calculated via rounds of multiplier effect. The final output and direct output jointly constitute the total output of the contribution of cruise tourism. The calculation of the multiplier effect of cruise tourism mainly applies Keynesian model method, input-output model method and Computable General Equilibrium (CGE) model method.

1.4.1 Keynesian Model

Since Keynesian model can calculate the value of the multiplier and the data and calculation method required are relatively simple, it is often used to conduct simple preliminary calculation of multiplier effect in case of incomplete data. The following data is a typical Keynesian model:

> The L in the numerator reflects the proportion of imported commodities and services. (1 − L) is the proportion of consumed commodities and services actually produced locally. The denominator reflects the proportion of consumer income not further used in local consumption (income leakage loss). If we assume that all commodities and services sold in a region are actually produced locally, i.e. L = 0, that the consumers do not import any commodity and service, i.e. MPM = 0; that the local government does not collect any tax, i.e. MTR = 0, that the local consumers save CNY 20 of each CNY 100 that they receive, i.e. MPS = 0.2, the Keynesian multiplier for the region can be calculated by the model, indicating that the local consumers' income of each CNY 1 can finally bring the region with income of CNY 5 by means of the multiplier effect. It can be concluded that the higher ratio of the amount that may be used for further consumption to local consumers' income, the stronger promoting function on follow-up industries, and the higher the Keynesian multiplier. On the contrary, if only a small portion of their income can be used for further income, the Keynesian multiplier is lower.

The data and calculative process required for Keynesian model are the simplest. The model is widely used for the evaluation on the economic contribution of the cruise industry in the small countries along the Caribbean Sea. However, this model evaluation method has many defects. First, Keynesian model applies partial equilibrium analysis without considering production substitution effect and other mechanisms. Compared with input-output model, CGE model and other general equilibrium analysis, theoretical assumption does not fully comply with the economic reality and may have great error in calculation results if the economic scale in the region studied is very large and the economic structure is complicated. Second, Keynesian model applies static analysis method, therefore cannot reflect the dynamic change of economic structure. Third, Keynesian model can only get the value of multiplier, but cannot reflect the internal relation among various industries, or obtain data on the total economic contribution of cruise industry by sectors. Therefore, some scholars held that "even the most complicated Keynesian model cannot meet the decision-making requirements of the government" (Vanhove 2011).

1.4.2 Input-Output Model

Input-output model is a static general equilibrium model and is the main method applied by various countries in the world to evaluate the multiplier effect and the total economic contribution of cruise industry tourism. Based on the analysis of regional input-output table, input-output model can reflect the interrelation among various industries and the actual internal structure of economic entities, reflect the valued added, intermediate inputs, intermediate products and the flow direction and the flow of final products, and calculate the direct output and resultant output driven by the multiplier effect. Input-output model can calculate not only the complete consumption by the direct cruise tourism consumption of the total output of economic entities, but also the complete consumption of the output of single industry and the values for the total economic contribution of cruise industry, such as output, value added, employment, income, profits and taxes. Government departments can, based on such data, formulate targeted industrial policies. Nevertheless, input-output model also has some demerits:

First, the returns to scale are assumed to be invariant, which cannot reflect the economies of scale in the process of production. That is to say, the inputs for producing 10 products are assumed to be 10 times of that for one product; and in many industries, the unit cost may be saved with the increase of output.

Second, it is assumed that there is perfectly elastic total supply curve, that is to say, all production factors required for output can be automatically met and the prices of production factors remain unchanged, causing no production substitution effect as mentioned above, and also can not impose any impact on other industries or other regions. This assumption is true if the scale of cruise industry is very small. However, when the annual cruise tourists reach 2 million person-times in Shanghai, it is difficult to guarantee that such large-scale cruise tourism consumption will not impose impact on the product price.

Third, the direct cruise tourism consumption is supposed to be an exogenous variable, which is not determined by the specific microscopic behavioral model under certain economic and social background; the substitution rate of marginal consumption is supposed to be a constant value, that is, the product price, resident income and other changes in the operation of cruise economy will not, in turn, influence cruise tourism consumption.

Fourth, the input-output table referred by the research institute is comparatively outdated and may not reflect the current production technology and economic structure. For example, the latest input-output table available to Shanghai is the *Input-output Table of China (2007)* released by the National Bureau of Statistics (NBS), the relevant data in which has been eight years from now and may differ from the economic reality of Shanghai.

Regardless of the above defects, input-output model is still the most practical and most widely used model in the evaluation on economic contribution all over the country. It is widely applied in the evaluation system for economic contribution in various regions.

1.4.3 CGE Model

CGE model, a DSGE model based on input-output model, covers resource constraints, production factor substitution, product price substitution, changes in demand and income and other aspects, and can reflect the dynamic changes in producer behavior, consumer behavior and industrial connection in the process of economic growth of cruise. The results therefrom will be more realistic and reasonable. Actually, CGE model lowers some of the strict prerequisite of assumption in input-output model, considers the complex interaction between production and consumption in the process of economic growth of cruise industry, and covers the comprehensive cost and negative effect of cruise tourism. Therefore, the economic contribution calculated by CEG model is generally lower than that by input-output model (Stabler et al. 2010). The calculation results of CGE model may even be negative if the negative effect is great.

CGE model is a most complete and detailed multiplier effect estimation method. With fewer restrictive assumptions for the operation of economic entities, it can fully reflect the motivation and behavior of producers and consumers and reflect economic realities more objectively. CGE model has been applied in the evaluation on tourism economy impact effect since 1990s, but it is rare in the evaluation on the impact effect of cruise economy. CGE model, established on the basis of input-output model, requires detailed survey data and complicated calculation model, and is gradually established when the input-output model gets very perfect.

2 Evaluation System for Economic Contribution of Cruise Industry to Shanghai

Based on the analysis of the theory, method and case for evaluation on the economic contribution of cruise industry, this chapter analyzes the characteristics and development trend of the cruise economy in Shanghai, and aims at establishing an evaluation system for economic contribution of cruise industry to Shanghai.

2.1 Characteristics and Trends of Cruise Economy in Shanghai

In order to design the evaluation system for economic contribution of cruise industry, the characteristics of the cruise economy in Shanghai should be first objectively analyzed. The characteristics and development trend of cruise economy in Shanghai can be summarized as follows:

2.1.1 Diversified Structure of Cruise Tourists

Under the background where the pilot reform experience in free trade pilot site is promoted in Baoshan District and the "integration of free trade zone and port" system innovation is actively promoted and explored, the convenience degree for inbound cruise tourists may be effectively improved, the proportion of foreign tourists in cruise tourists increased and the number of cruises with Shanghai as the port of call by implementing Shanghai Air Harbor 72 h visa- free transit policy at Shanghai Wusongkou International Cruise Terminal, exploring to establish harbor and airport linkage mechanism for overseas tourists, carrying out the corresponding entry facilitation measures, and keeping optimizing the supervision service environment at cruise port.

With the enlargement of domestic tourism market for the cruise tourism in Shanghai and the gradual maturity of the development conditions for offshore cruise tourism business, the ratio of long-distance tourists to the cruise tourists in Shanghai will keep increasing, together with that of tourists who stay at home port and port of call tourists, which will change the situation where tourists who do not stay at home port account for an overwhelming majority.

2.1.2 Great Importance Attached to Ship Supply Industry

Ship supply industry is the most important part of cruise industry and the critical function of cruise home port, accounting for 72.1% of the expenditures of cruise industry among the direct economic contribution of cruise industry all over the world.

Due to tax, customs inspection and other systems, most of cruise enterprises operating in Northeast Asian choose to establish cruise material distribution center in Pusan, Korea. Sometimes, such enterprises even transport the commodities from China to Pusan and then provide such commodities to cruises with home port in China. As the largest cruise home port in Asia, Wusongkou now only distributes some fresh food for cruise, which is highly inconsistent with its position in industry. With the promotion of the innovation of customs supervision system and the system of inspection and quarantine and the gradual maturity of domestic and overseas material distribution mode, the establishment of the special bonded warehouse for ship supply materials and cruise material distribution center in Shanghai will make a breakthrough. The ship supply industry will develop rapidly.

2.1.3 Gradual Extension of Cruise Services

Shanghai is actively exploring and developing cruise maintenance business and implementing the global inspection and quarantine supervision system in maintenance industry to promote the operation of cruise outbound tourism business by travel agency registered in the region. Meanwhile, Shanghai actively establishes cruise talent education training institutions and vocational training institutions to strengthen the training of R&D and sales personnel and lead guides and establish cruise talent intermediary. It can be seen that the range of services for cruise industry in Shanghai is developing rapidly, making more and more contribution to the economic growth in Shanghai.

2.1.4 Cruise Headquarters Economy

With the rapid expansion of the cruise economy in China, more and more large international cruise enterprises choose to establish regional headquarters in Shanghai. Many domestic cruise enterprises also establish headquarters in Shanghai. Cruise headquarters economy has begun to take shape. With the further improvement of the business environment in Shanghai for cruise enterprises and the further enhancement of the agglomeration effect of cruise headquarters economy, a large number of enterprises become a constituent part of Shanghai cruise headquarters base. Besides, Shanghai will also be the preferred location of international cruise organization or regional headquarters for industry association, which will further the function of cruise headquarters economy.

2.1.5 Rapid Development of Cruise Supporting Industries

Shanghai actively develops such industries as cruise port service, cruise supporting services, land and water tourism, shipping agents, cruise equipment technology research, cruise design and building, cross-border commodity trading center and electric business platform for cruise tourism around the cruise port. The industry chain of cruise is further extended, the coverage of cruise industry is further expended and its promotion to the whole economy is further enhanced.

2.1.6 Breakthrough in Local Cruise Industry

Shanghai actively cultivate and establish local cruise operating enterprises, set up local cruise companies in Shanghai and encourage international cruise companies to establish joint cruise tourism enterprises in Shanghai. With the first navigation of "Tianhai Golden Era" Cruise at Shanghai Wusongkou International Cruise Terminal on May 15, 2015, the local cruise industry in Shanghai has made great breakthrough. With the rapid expansion of domestic cruise market, it can be forecasted that the local cruise industry in Shanghai will develop rapidly. Compared with foreign cruise enterprises, local cruise enterprises can reserve more operating surplus in Shanghai for consumption or investment. More foreign commodities and services can be procured, more local employees engaged, and more taxes paid locally. Therefore, local cruise industry can bring Shanghai more economic effect.

2.1.7 Further Improvement of Public Services for Cruise

Shanghai Wusongkou International Cruise Terminal and Beiwaitan International Passenger Transportation Center keep improving infrastructure construction, enhance the cruise port shore power technology, and target at realizing the upgrade and expansion of infrastructure in combination with the subsequent engineering construction of Shanghai Wusongkou International Cruise Terminal. Meanwhile, Shanghai tries to improve the comprehensive service management level in cruise port region, optimize traffic, security and service facilities, and improve the environment around cruise port. It can be seen that with the rapid development of cruise industry, the public services of cruise will be rapidly improved in quality and quantity; and the amount of government consumption expenditures will also keep increasing; government consumption will be the important sources of the economic contribution of cruise industry.

2.1.8 Enhancement of Ability to Develop Urban Tourism

Cruise tourism has been an important driver for the rapid development of urban tourism in Shanghai. For example, the travel agencies, tourist hotels and tourist attractions in Baoshan District received 9.8164 million domestic and overseas tourists and had tourist income of CNY 1.308 billion in 2014, taking a leap in tourism. With the construction of tourist destinations in Shanghai, the complementarity between cruise tourism and urban tourism is further enhanced, indicating that cruise tourism will play a greater role in developing urban tourism.

2.2 The Economic Contribution of Cruise Industry to Shanghai

We can find that the economic contribution of cruise industry to Shanghai is comprehensive and profound by analyzing the characteristics and development trend of cruise economy in Shanghai. For example, Baoshan District treats cruise economy development as the strategic pillar by constructing Shanghai Wusongkou International Cruise Terminal, and cultivates and extends cruise industry chain by integrating various resources to realize the development path from "cruise terminal" to "cruise port" and then to "cruise city". Therefore, a comprehensive indicator system should be applied to evaluate the economic contribution of cruise industry to Shanghai to fully reflect the economic effect of cruise. Upon comprehensive consideration of the particularity of Shanghai cruise economy, the economic contribution of cruise industry has the following six major sources (Fig. 1).

2.2.1 Consumption of Cruise Tourists

It refers to all tourism consumption in Shanghai generated by tourists due to cruise tourism, including the consumption at the port and the terminal, the consumption during urban tourism in Shanghai before or after cruise tourism, and the transportation fee by inter-provincial transportation from and to Shanghai incurred by

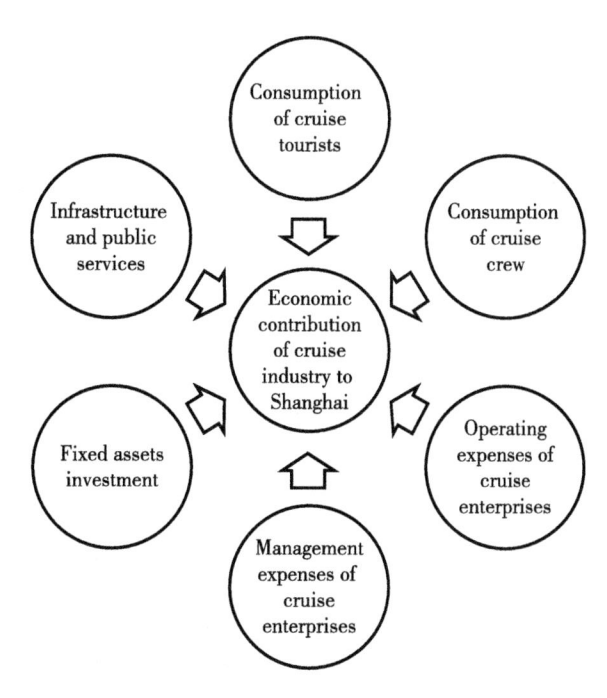

Fig. 1 Economic contribution of cruise industry to Shanghai

nonlocal tourists. The consumption of cruise tourists can be generally divided into long-distance transportation, urban transport, accommodation, catering, shopping, urban sightseeing, entertainment, communication and other categories.

It should be noted that the consumption does not include cruise ticket and the consumption on cruise by tourists. The total inter-provincial transportation fee will be only included in the income of transportation enterprises in Shanghai, and shall be determined by the means of transportation, such as HSR, airplane, long-distance bus and self-driving. In addition, such factors as the registration places of different transportation enterprises should be taken into account.

In order to accurately calculate the consumption data of cruise tourists in Shanghai, the two indicators, the number of cruise tourists and the consumption per person are required:

Generally speaking, the statistics about the number of cruise tourists are relatively comprehensive and can be directly obtained from cruise port, Shanghai Municipal Office for Port Services, Shanghai General Station of Immigration Inspection and other departments. However, tourists who stay at home port, tourists who do not stay at home port and port of call tourists should be distinguished. Generally, tourists who do not stay at home port and port of call tourists do not have any accommodation consumption; and port of call tourists do not have long-distance transportation fee. Tourists can be directly divided into home port tourists and port of call tourists according to the cruise navigation plan. Such factors as the origin of trip and the consumer behaviors of tourists should be considered when determining whether tourists stay overnight in shanghai. The demographic characteristics of tourists, such as nationality, gender and age, also should be recorded to study the consumer behaviors of different cruise tourists.

The consumption per person of cruise tourists is mainly obtained by the survey on tourist consumption behaviors. For group cruise tourists, partial data on consumption ashore can be obtained from travel agencies; the data on the consumption of cruise tourists in cruise passenger terminal building and in the port area can be obtained from relevant tourism enterprises. The inter-provincial transportation fee can be calculated by the origin of trip.

2.2.2 Consumption of Cruise Crew

It refers to the relevant consumption occurs in Shanghai by cruise crew, including the consumption in Shanghai during short stop of cruise and the consumption in Shanghai before and after the operation of cruise, but their consumption on the cruise is not included. Similar to the consumption of cruise tourists, cruise crew can also be divided into long-distance transportation, urban transport, catering, shopping, urban sightseeing, entertainment, communication and other categories. Generally, they do not have any accommodation fee.

Two indicators, the number of cruise crews who go ashore and their consumption per person, are required for the calculation of the consumption of cruise crew. The number of cruise crews who go ashore can be obtained from such departments as cruise port, Shanghai Municipal Office for Port Services and Shanghai General Station of Immigration Inspection. Cruise crews can be directly divided into home port crew and port of call crew according to the cruise navigation plan.

The consumption per person of cruise crew is mainly obtained by the survey on their consumption behaviors. If certain consumption ashore is incurred by cruise due to business reasons such as transfer at a port, relevant data can be obtained from cruise enterprises. In the process of questionnaire survey, cruise crews can be divided by whether Shanghai is their place of residence. If not, their consumption ashore shall be surveyed; if so, all income of crews can be included as consumption ashore since they live in Shanghai for long term.

2.2.3 Operating Expenses of Cruise Enterprises

It refers to the expenditures of cruise enterprises for cruise operation and onboard service relevant to Shanghai, including ship supply cost, fuel cost, cruise maintenance cost, port service cost and other costs, but the management expenses relevant to office ashore in Shanghai are not covered. Since cruise enterprises have global ship supply system, it should be noted that only the products and services purchased locally in Shanghai can be calculated.

Since cruise enterprises have kept detailed financial record, relevant data may be directly obtained from cruise enterprises. If cruise enterprises refuse to provide relevant data for the sake of trade secrets and other reasons, such data may be obtained from ship supply enterprises, fuel supplying enterprises, cruise maintenance enterprises, port service enterprises and other entities, or from shipping agents.

2.2.4 Management Expenses of Cruise Enterprises

The management expenses incurred from the establishment of headquarters or branches in Shanghai by cruise enterprises, generally including the remuneration of office staff, enterprise office expenses, marketing cost, service outsourcing cost and other costs.

Generally, most of office staff live in Shanghai for a long time. Therefore, in order to simplify the calculation process, the total remuneration of office staff should be included in the local consumption of residents in Shanghai. The specific consumption structure may refer to the structure of consumption expenditure of urban households in the *Shanghai Statistical Yearbook*.

Enterprise office expenses include the rent, office allowance, communication cost, property cost and other costs borne by cruise enterprises in Shanghai.

Marketing cost includes the travel expenses, transportation, advertising costs, agency fee, packing expenses and other expenditures incurred by cruise enterprises in Shanghai for the sales of products. The costs for marketing in other cities than Shanghai should be excluded.

Service outsourcing cost refers to the expenses for outsourcing partial management services of cruise enterprises to the enterprises in Shanghai. The cost for outsourcing to enterprises beyond Shanghai should be excluded.

The above data may be obtained by survey on cruise enterprises, check of annual reports of cruise companies, approximate enterprise calculation and other ways.

2.2.5 Fixed-Assets Investment

First, fixed assets investment includes the costs for purchasing cruise built in Shanghai. Since cruise is the most important fixed asset of cruise enterprises, the costs should be gradually amortized in the process of the operation of cruise enterprises for years. Therefore, the costs are included in the fixed assets investment of cruise enterprises, rather than the operating expenses of such enterprises. Shanghai has not made any major breakthrough in the building of cruise, thus cruise shipbuilding industry makes little economic contribution. When such breakthrough is made in the future, relevant data can be obtained from cruise enterprises, or from cruise building enterprises. It should be noted that no matter whether Shanghai is the home port of the cruises built in Shanghai, cruise building costs should be included in the fixed assets investment of cruise enterprises.

Second, the expenditures of cruise enterprises, cruise organizations, cruise industry associations and other entities for purchasing vehicles and office space or constructing office buildings can also drive the economic growth in Shanghai in the form of investment, and should be included in cruise-related fixed assets investment. Investment data can be directly collected from cruise enterprises or cruise organizations or by checking the annual reports of enterprises or organizations.

2.2.6 Infrastructure and Public Services

First, in order to promote the development of cruise tourism, the governments at all level in Shanghai constructed the transport infrastructure, landscape facilities, public tourism scenic spots, leisure and entertainment facilities, tourist service center, tourist toilets and other tourism service facilities for the construction of cruise port. Even though such facilities cannot acquire direct income from tourists, crews or cruise enterprises, the economic growth in Shanghai can be driven in the process of construction in the form of investment. Therefore, relevant costs should be included in cruise-relevant infrastructure facility investment.

Second, cruise-relevant infrastructure facilities still require huge expenditures in the operation and maintenance of such facilities as roads, parks, museums and toilets. The daily operation of such facilities can also promote the consumption of government and increase the economic growth in Shanghai. Therefore, relevant costs should be included in cruise-relevant public service consumption.

The data on the investment into cruise-relevant infrastructures and the expenditures for public services can be obtained from such departments as development

Table 4 Data acquisition method for evaluation on economic contribution of cruise industry to Shanghai

Category	Indicator	Data source	Remarks
Economic and social background data	Labor productivity by sectors	1. Municipal Statistics Bureau 2. Data from the third economic census	The data should be detailed on medium class and subclass
	Input-output table	1. Municipal Statistics Bureau	The latest input-output table after 2012 and with more than 135 departments should be applied
	Inter-provincial trade	1. Municipal Statistics Bureau 2. Municipal Commerce Commission	Proportion of cruise-relevant commodities and services from other regions than Shanghai should be calculated
	Foreign trade	1. Municipal Statistics Bureau 2. Municipal Commerce Commission	Proportion of cruise-relevant overseas commodities and services should be calculated
Consumption of cruise tourists	Person-times of cruise tourists	1. Cruise port 2. Shanghai General Station of Immigration Inspection	Tourists are divided into home port tourists and port of call tourists according to the cruise navigation plan
Consumption of cruise tourists	Structure of cruise tourists	1. Travel agency 2. Shanghai General Station of Immigration Inspection	The demographic characteristics of cruise tourists, such as nationality, gender and age, should be recorded
	Structure of home port tourist sources	1. Travel agency 2. Shanghai General Station of Immigration Inspection 3. Survey on tourist consumption behaviors	The structure of home port tourist sources should be recorded. Home port tourists are divided into tourists who stay at home port and tourists who do not stay at home port
	Period for tourists to stay in ports of call	1. Travel agency 2. Cruise enterprises 3. Shanghai General Station of Immigration Inspection 4. Survey on tourist consumption behaviors	The period for tourists to stay in Shanghai as a port should be calculated to estimate the consumption
	Consumption per tourist for shore tourism	1. Travel agency 2. Cruise passenger terminal building 3. Cruise-relevant tourism enterprises 4. Survey on tourist consumption behaviors	All kinds of consumption of cruise tourists in Shanghai should be included

(continued)

Table 4 (continued)

Category	Indicator	Data source	Remarks
Consumption of cruise crew	person-times of cruise crew who go ashore	1. Cruise port 2. Shanghai General Station of Immigration Inspection	Crew is subdivided into home port crew and port-of-call crew
	Structure of cruise crew	1. Cruise enterprises 2. Shanghai General Station of Immigration Inspection	The demographic characteristics of cruise crew, such as nationality, gender and age, should also be recorded
	Structure of home port crew sources	1. Cruise enterprises 2. Shanghai General Station of Immigration Inspection 3. Survey on crew consumption behaviors	The structure of home port crew sources is analyzed. Home port crews are divided into two kinds by whether Shanghai is their place of residence
	Period for screw to stay ashore in ports of call	1. Cruise enterprises 2. Shanghai General Station of Immigration Inspection 3. Survey on crew consumption behaviors	The period for screw to stay in Shanghai as a port of call should be calculated to estimate the consumption
Consumption of cruise crew	Consumption per crew member ashore	1. Cruise passenger terminal building 2. Crew service enterprises 3. Survey on crew consumption behaviors	All kinds of consumption of cruise crew in Shanghai should be included
Operating expenses of cruise enterprises	Ship supply cost	1. Cruise enterprises 2. Ship supply enterprises	Such ship supply materials as food, beverage, daily necessities and consumables purchased in Shanghai by cruise enterprises should be included
	Fuel cost	1. Cruise enterprises 2. Fuel oil supplying enterprise 3. Shore power supplying enterprise	The fuel oil and shore power purchased in Shanghai by cruise enterprises should be included
	Cruise maintenance cost	1. Cruise enterprises 2. Cruise maintenance enterprises 3. Cruise decoration and rebuilding enterprises	Cruise-relevant daily maintenance costs and decoration and rebuilding costs incurred in Shanghai should be included
	Port service cost	1. Cruise enterprises 2. Cruise port 3. Port service enterprises	Shanghai cruise port charge for use and other port service charge, such as the cost of tug, navigation, berthing, loading and unloading should be covered

(continued)

7 The Economic Contributions of Evaluation Indicator System ...

Table 4 (continued)

Category	Indicator	Data source	Remarks
Management expenses of cruise enterprises	Remuneration of office staff	1. Cruise enterprises 2. Annual reports of companies 3. Municipal Local Taxation Bureau	The remuneration of office staff should be regarded as the consumption of residents in Shanghai
	Enterprise office expenses	1. Cruise enterprises 2. Annual reports of companies 3. Approximate enterprise calculation	The rent, office allowance, business entertainment cost, travel expenses, repair cost, communication cost, vehicle cost, R&D cost, property cost and other costs borne by cruise enterprises in Shanghai for office ashore should be covered. Only the enterprise office expenses incurred in Shanghai should be included
Management expenses of cruise enterprises	Marketing cost	1. Cruise enterprises 2. Annual reports of companies	Marketing cost includes the travel expenses, transportation, advertising and general publicity expense, agency fee, packing expenses and other expenditures incurred by cruise enterprises in Shanghai for the sales of products. The costs for marketing in other cities than Shanghai should be excluded
	Service outsourcing cost	1. Cruise enterprises 2. Annual reports of companies	It refers to the expenses for outsourcing partial management services of cruise enterprises, such as training, to the enterprises in Shanghai
Fixed assets investment	Cruise purchasing cost	1. Cruise enterprises 2. Shipbuilding enterprise	It only covers the cost for purchasing cruise built in Shanghai
	Fixed assets investment expenditures	1. Cruise enterprises 2. Cruise organization 3. Cruise Industry Association 4. Annual reports of companies 5. Municipal Urban and Rural Construction and Management Committee	This covers the expenditures of cruise enterprises, cruise organizations, cruise industry associations and other entities for purchasing vehicles and office space or constructing office buildings

(continued)

Table 4 (continued)

Category	Indicator	Data source	Remarks
Infrastructure and public services	Infrastructure investment	1. Municipal Development and Reform Commission 2. Municipal Urban and Rural Construction and Management Committee 3. Municipal finance bureau	It covers the investment into cruise- relevant transport infrastructure, landscape facilities, public tourism scenic spots, leisure and entertainment facilities, tourist service center, tourist toilets and other tourism service facilities
	Public service expenditures	1. Municipal Finance Bureau 2. Municipal greening and administration of city appearance 3. Municipal Cultural Administration of Radio, Film and Television and Tourism Bureau 4. Municipal Tourism Bureau	This covers the expenditures in the operation and maintenance of such facilities as roads, parks, museums and toilets

and reform commission, finance bureau, construction committee, cultural administration of radio, film and television and tourism bureau. It should be noted that this item only calculates the infrastructure and public services, for which cruise enterprises and tourists in Shanghai do not have to pay. All facilities at cruise port generally charge for use, which has been calculated in the operating expenses of cruise enterprises. Therefore, such charge will not be included in this item to avoid repetitive computation.

2.3 Data Acquisition Method for Evaluating Economic Contribution of Cruise Industry to Shanghai

The main sources of economic contribution of cruise industry to Shanghai are consumption of cruise tourists, consumption of cruise crew, operating expenses of cruise enterprises, management expenses of cruise enterprises, fixed assets investment, and infrastructure and public services. In order to calculate the contribution of each part to the economy of Shanghai, an indicator system should be established based on the characteristics of each part and evaluation data collected from multiple sources.

Table 4 lists the evaluation indicator system and data acquisition method for the six major sources of the economic contribution of cruise industry to Shanghai, and states the problems to which attention should be paid in the process of data collection. Since the cruise industry in Shanghai appeared relatively late and was defined as consumption complementary, it is difficult to obtain relevant data directly from statistical yearbook or official report. Therefore, a series of targeted data collecting methods are required to effectively evaluate the economic contribution of cruise industry. Specifically, the data in the following 10 aspects are required for the comprehensive evaluation on the economic contribution of cruise industry to Shanghai:

- Statistical data from the *Shanghai Statistical Yearbook* or the third economic census
- Survey on the development and reform commission, commerce commission, finance bureau, tourism bureau and other government departments
- The exit and entry statistics from Shanghai Municipal Office for Port Services and Shanghai General Station of Immigration Inspection
- Survey on cruise enterprises
- Survey on cruise port and cruise services enterprises
- Survey on cruise passenger terminal building and tourism enterprises around the port
- Survey on cruise travel agencies
- Survey on consumption behaviors of cruise tourists
- Survey on consumption behaviors of cruise crew
- Survey on cruise organizations, cruise industry associations and other NGOs.

2.4 Evaluation Indicator System for Economic Contribution of Cruise Industry to Shanghai

2.4.1 Technical Route for Evaluation on Economic Contribution of Cruise Industry to Shanghai

Upon the collection of complete data on the detailed indicators of the six major sources for the economic contribution of cruise industry to Shanghai, such data should be processed and sorted to calculate the direct economic contribution and total economic contribution of the cruise industry to Shanghai.

Figure 2 lists the detailed technical route for the evaluation on the economic contribution of cruise industry to Shanghai and showed the whole process where the direct economic contribution and total economic contribution of cruise industry are calculated on the basis of the data for the evaluation on the economic contribution of cruise industry.

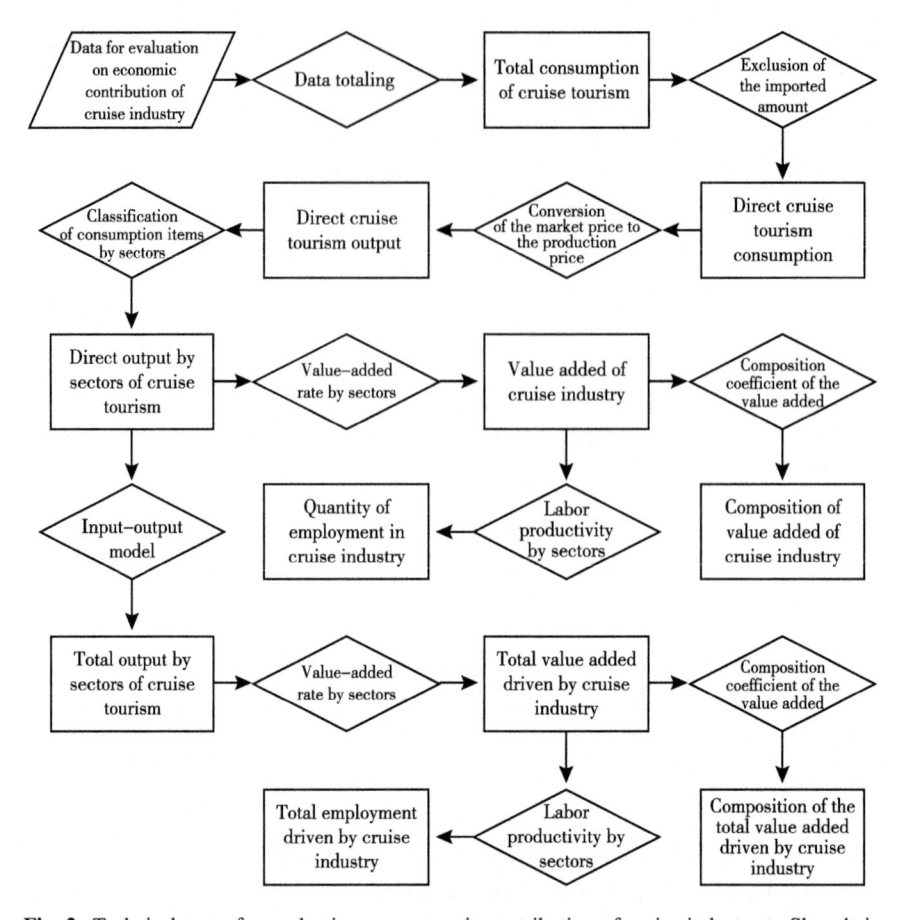

Fig. 2 Technical route for evaluation on economic contribution of cruise industry to Shanghai

2.4.2 Indicators for Evaluation on Economic Contribution of Cruise Industry to Shanghai

A comprehensive indicator system should be established to evaluate the economic contribution of cruise industry to Shanghai. The indicators can be divided into the "direct economic contribution" and the "total economic contribution". See Table 5 for the specific indicators and its connotation.

2.4.3 Method to Calculate Indicators for Direct Economic Contribution of Cruise Industry to Shanghai

In order to evaluate the direct economic contribution of cruise industry to Shanghai in a more comprehensive manner, the two indicators, the direct cruise tourism

Table 5 Evaluation indicators for economic contribution of cruise industry to Shanghai

Category	Indicator	Connotation
Direct economic contribution	Direct cruise tourism consumption	The expansion of the direct output of cruise industry in Shanghai, including the value added and intermediate inputs in cruise industry
	Value added of cruise industry	The development of cruise industry in Shanghai
	Laborers' remuneration	The remuneration of labors in cruise industry in Shanghai
	Net production tax	The tax paid by cruise enterprises in Shanghai
	Depreciation of fixed assets	Depreciation of fixed assets of cruise enterprises in Shanghai
	Operating surplus	The profits gained by cruise enterprises in Shanghai
	Quantity of employment in cruise industry	The total quantity of employment in the cruise industry in Shanghai
Total economic contribution	Total cruise tourism output	The expansion of the total output in various industries brought about by cruise tourism in Shanghai
	Total value added driven by cruise	The expansion of the total scale in various industries brought about by cruise tourism in Shanghai
	Laborers' remuneration	The increased laborers' remuneration in various industries brought about by cruise tourism
	Net production tax	The increased total tax paid by enterprises in various industries driven by cruise tourism in Shanghai
	Depreciation of fixed assets	The increased total depreciation of fixed assets of enterprises in various industries driven by cruise tourism in Shanghai
	Operating surplus	The increased total profits gained by enterprises in various industries driven by cruise tourism in Shanghai
	Total quantity of employment driven by cruise	The increased total quantity of employment in various industries driven by cruise tourism in Shanghai

consumption and the value added of cruise industry in Shanghai, should be calculated. The direct cruise tourism consumption can reflect the direct economy driving in Shanghai by cruise tourism and is an important indicator for evaluation on the direct economic contribution of cruise industry. It includes the value added and intermediate inputs in cruise industry. Therefore, its value is generally higher than the value added of cruise industry. The value added of cruise industry can reflect the development of the cruise industry in Shanghai and can be used to calculate the proportion of cruise industry in the GDP of Shanghai or compare cruise industry with other industries in Shanghai.

(1) Direct cruise tourism consumption

Upon the acquisition of data for the evaluation on economic contribution of cruise industry, the total cruise tourism consumption in Shanghai can be calculated by totaling the consumption and investment of the six major sources for the economic contribution of cruise industry. However, it should be noted that this is not the real direct economy driving in Shanghai by cruise tourism. The reason is that even such consumption and investment are generated in Shanghai, the commodities and services purchased may be from outside of the city or even overseas and this part has little driving function on local economy. The imported ratio in each industry can be calculated by checking the latest input-output table or by the survey on such departments as the municipal commerce commission to exclude this part from the total cruise tourism consumption to calculate the direct cruise tourism consumption in Shanghai.

(2) Value added of cruise industry

First, the direct cruise tourism consumption in Shanghai reflects the market price of the commodities and services consumed by cruise tourism in Shanghai. Compared with the production price, the market price additionally includes dealer margins and circulation costs. Each sector in cruise industry can be actually reflected and direct cruise tourism output obtained when the market price is reduced to the production price by deducting such intermediate costs.

Second, since the consumption data collected from the questionnaire surveys are classified by consumption items, when calculating the promotion of the economy by cruise tourism consumption, the consumption data should be classified into corresponding industries of national economy by the nature of products to get the direct output of various sectors of cruise industry. For example, if it is known, by survey on cruise enterprises, that a cruise enterprise purchased the office furniture valued CNY 1 million and manufactured in Shanghai, such CNY 1 million should be classified into "timber processing and furniture manufacturing" according to the *Industrial classification for national economic activities*.

Third, when the direct output of sectors of cruise tourism is calculated, the value-added rate of each sector, i.e. the ratio of the value added in each sector, can be calculated by checking the latest input-output table to calculate the value added in each sector of cruise industry and the value added of cruise industry by totaling such values.

Fourth, when the value added of each sector of cruise industry is calculated, the laborers' remuneration coefficient, net production tax coefficient, coefficient of depreciation of fixed assets and coefficient of operating surplus can be calculated according to the input-output table to further divide the value added of cruise industry in Shanghai into the laborers' remuneration, net production tax, depreciation of fixed assets and operating surplus, which actually reflect the distribution of the value added of cruise industry among the employees of enterprises, government and enterprise owners.

Finally, the quantity of employment in cruise industry can be obtained by backward calculation on the basis of the labor productivity in various industries in Shanghai and the value added by sectors of cruise industry. The calculated quantity of employment can only be the Full Time Equivalent rather than the actual quantity of employment for cruise industry, and labors may work overtime in rush season and be employed in other industries in off-season.

2.4.4 Method for Calculating Indicators of Economic Contribution of Cruise Industry to Shanghai

The above indicators of the direct economic contribution of cruise industry, such as the direct cruise tourism consumption and the value added of cruise industry, can only reflect the direct economy driving function of cruise tourism in Shanghai. Upon the direct driving of the production in various industries in Shanghai by cruise industry, such industries can drive the production in intermediate input industries. The direct output and resultant output can be calculated via rounds of multiplier effect. The final output and direct output jointly constitute the total output of the contribution of cruise tourism.

(1) Comparison of different evaluation models

Three models, Keynesian model, input-output model and CGE model, are available to the evaluation on the total economic contribution of cruise industry to Shanghai. Keynesian model cannot reflect the economic relation among industries and cannot get the economic contribution indicators of sectors, always leading to great error in the evaluation results. This model cannot meet the decision-making consultation requirements of Shanghai Municipal People's Government. Therefore, it is not applied herein. CGE model is a DSGE model based on input-output model and has higher requirements for the comprehensiveness and accuracy of evaluation data. It cannot meet the data requirement at the initial stage of the establishment of the evaluation indicator system for economic contribution of cruise industry to Shanghai, thus it is not used herein. CGE model may be applied when such system gets mature and perfect. Therefore, according to international practice, the input-output model is mainly applied to evaluate the total economic contribution of cruise industry to Shanghai.

(2) Input-output model

Since the direct cruise tourism output by sectors was calculated in the above process of the evaluation on the direct economic contribution of cruise industry, the total cruise tourism output can be obtained by calculating the matrix of total consumption coefficient in the input-output table and multiplying by the column vector of the direct cruise tourism output. The total cruise tourism output reflected the expansion of the total output in various industries brought about by cruise tourism in Shanghai.

When the total cruise tourism output is calculated, the total value added by sectors in Shanghai driven by cruise tourism can be calculated by checking the input-output table. It reflects the expansion of the total scale in various industries brought about by cruise tourism in Shanghai.

When the total value added by sectors in Shanghai driven by cruise tourism is calculated, the total value added by sectors can be divided into laborers' remuneration, net production tax, depreciation of fixed assets and operating surplus according to the laborers' remuneration coefficient, net production tax coefficient, coefficient of depreciation of fixed assets and coefficient of operating surplus in the input-output table. Such indicators reflects the total increased income that cruise industry can bring about to the employees of enterprises, government, enterprise owners and other main bodies in Shanghai.

When the total value added by sectors in Shanghai driven by cruise tourism is calculated, the increased total quantity of employment in various industries driven by cruise tourism in Shanghai can be obtained by backward calculation on the basis of the labor productivity in each industry. This indicator reflects the increased total quantity of employment in various industries driven by cruise tourism in Shanghai.

2.5 Summary

This study establishes the evaluation method and evaluation indicator system for the economic contribution of cruise industry to Shanghai. However, it is still difficult to provide a full view of such economic contribution due to the limited data. Nevertheless, seen from the conclusion of the analysis, the influence of cruise economy on the economic development of Shanghai is profound. Cruise industry can be an important mainstay industry in economy. With the rapid expansion of the number of cruise tourists, continuous increase of the proportion of tourists who stay at home port, the general improvement of cruise supplying industry and the gradual formation of cruise headquarters economy, cruise economy will rapidly expand accordingly and play a more important role in the economic development of Shanghai.

Chapter 8
Development Path Research on Shanghai's Cruise Supporting Industry

Guodong Yan, Shuguang Lei, Yanhui Gao, Ying Ye and Yanan Cai

1 Research Background

1.1 China Has Stepped into Economic New Normal, and It Is a Significant Opportunity for Transformation and Development

According to the *13th Five-year Tourism Development Planning*, we should vigorously develop marine and waterfront tourism, take developing cruise tourism planning as key point, pay attention to basic equipment and matching condition required for developing and improving cruise tourism and constantly improve

---

Guodong Yan—Research direction: Cruise industry and Economic management.
Yanan Cai—Research direction: Cruise tourism personnel training.

G. Yan (✉)
Shanghai International Cruise Business Institute,
Shanghai University of Engineering Science, Shanghai, China
e-mail: yanguodong@163.com

S. Lei
Shanghai Baoshan Industrial Park, Baoshan District, China

Y. Gao
Shanghai International Cruise Tourism Service Center Co. Ltd.,
Shanghai Wusongkou International Cruise Port Development Co. Ltd.,
Shanghai, China

Y. Ye
Shanghai Baoshan District Economic and Information Work Committee,
Shanghai, China

Y. Cai
Shanghai University of Engineering Science, Shanghai, China

© Social Sciences Academic Press and Springer Nature Singapore Pte Ltd. 2018
H. Wang (ed.), *Report on China's Cruise Industry*,
https://doi.org/10.1007/978-981-10-8165-1_8

technological content and innovativeness of port. China has put forward Yangtze River Economic Belt and the 21st Century Maritime Silk Road construction, and Shanghai, as key node city for the two major national strategic deployments, should give full play to window and leading role; as a result, Shanghai Wusongkou International Cruise Terminal will be more strategically significant: It will become cruise ship economic core and experiment base and give play to its window function of marine economy. Currently, Baoshan District of Shanghai is in the key stage for innovation and transformation, industrial adjustment and urban upgrading and faced with a major opportunity: It is planning to create modern service industry and tourism and leisure clustering area in the north of Shanghai and leads functional optimization and industrial structural adjustment of riverside area and even the whole functional area.

1.2 Great Opportunity for Constructing Shanghai's World's Famous Tourism City and "Four Centers"

According to the 13th Five-year Planning for Shanghai's Tourism Reform and Development, it is required to promote development of new tourism format; of which, cruise tourism should strengthen overall planning, establish regional standard, optimize customs clearance policy and meet demand for cruise tourism development; improve industry chain for cruise tourism development and ensure Shanghai can supply materials to cruise ship enterprises nearby; dock internal service specification and actively win support of national relevant ministries and commissions; cultivate local cruise ship company and cruise ship fleet, attract more cruise ship enterprise headquarters and institutions to gather in Shanghai, establish cruise tourism distribution hub of Asian-Pacific region and accelerate and boost independent R&D and manufacturing of domestic cruise ship; establish domestic cruise ship professional supporting system and try to elementarily establish international cruise ship hub port of Asian-Pacific region and world's cruise ship and tourism core area at the end of the 13th Five-year Plan period. Shanghai should actively participate in construction of Shanghai International Shipping Center and offer necessary carrier support for expanding Shanghai's and even domestic shipping economic development connotation and optimizing cruise supporting industry development.

1.3 Optimization and Upgrading of Cruise Supporting Industry Accelerates and Improves Shanghai's Ability for Blending into Value Chain of Global Cruise Industry

Optimization and upgrading of China's cruise industry has been formed; but academic circle's research on cruise supporting industry is at starting stage, and there has not been a complete theoretical system and it is hard to provide effective theoretical support for development of cruise industry. This paper is based on development and research of Shanghai's cruise supporting industry to analyze and organize features, bottlenecks, improvement countermeasures and key measures of Shanghai's cruise supporting industry development through expert interview and positive analysis, which enriches research connotation of supporting industry to some degree, conforms to strategic requirement for China's cruise industry and tourism development and offers new research idea for accelerating overall linking and upgrading of Shanghai's cruise industry chain and quickening and improving Shanghai's overall capacity for blending into value chain of global cruise industry as well as reference for formulation of relevant policies and systems of cruise industry.

2 Literature Review

2.1 Existing Research on Development of Cruise Supporting Industry

2.1.1 Foreign Research

Foreign research on cruise industry originates from 1990s. In terms of research filed, it undergoes the process from macro to micro and is mainly focused on cruise tourism, cruise ship enterprise operation and cruise ship public supporting service. In terms of research content, foreign research on cruise industry is gradually deepened and expanded. To be specific, in research on cruise ship supply system, most of literatures are focused on regions where cruise industry is developed, such as Mediterranean region and Caribbean Sea region. The representative two articles about cruise ship supply system respectively take Caribbean Sea region and Greek Piraeus for instance. The former is concentrated on enterprise micro perspective and the latter is mainly oriented to function of cruise ship supply for local economy and makes a positive analysis. As for cruise tourism and service operation, James Mak and other researchers are focused on relationship between passenger ship service behavior and the US's cruise tourism. Simon Véronneau and Jacques Roy organize problems and challenges that cruise ship operation and management company faces. In the aspect of cruise ship home port and industrial economic research, in *2011 CLIA Cruise Market Overview*, market development and capacity of cruise ship terminals in the US, Canada and other countries are organized. Florida Harbor Transportation and Economic Development Commission in the US made a five-year planning for cruise ship system in the whole state in 2013; London Development Agency assessed facility demand of London's future cruise industry and had a case analysis for development of Amsterdam, Hamburg, Newcastle, Southampton, Dover, Harwich and other cruise ship terminal cities. New York City Economic Development Corporation (NYCEDC) issues research report for economic effect of New York's cruise industry on a yearly basis. In the aspect of cruise ship terminal industrial cluster, according to research of professional scholars in Belgium, the US, Netherlands, Norway, etc., port industrial cluster has a remarkable influence on employment rate and national economy.

2.1.2 Domestic Research

China's cruise industry is still at initial development stage. Relevant research gives corresponding opinions mostly from a macro perspective, and in-depth research is still lacking and mainly focused on aspects below: In terms of research territorial scope, it has an overall discussion on development of China's cruise industry and the positive research on development of a certain port. In *Development Report on China's Cruise Industry (2014–2016)* written by Wang Hong, etc., development of China's cruise industry is systematically analyzed. Wu et al. (2014) have conducted research on the development of China's cruise industry. Miao et al. (2013) analyzed factors influencing China's cruise industry development. With respect to research method, the research is mainly relied on qualitative research and only a small number of researches are completed through quantitative research, e.g. Qi et al. (2010) had measurement and analysis for competitiveness of cruise industry in China's port cities; Cai et al. (2010) had a measurement for development potential of China's cruise industry. With respect to research content, in the aspect of cruise industry chain research, Gan (2013) analyzed features of Shanghai's cruise industry chain; Sun (2015) analyzed relevant policies of cruise industry from the perspective of tourism science and industrial economics; Zhang (2014) put forward policy suggestions including loosening ship age restriction for cruise ship import and proportion restriction of Chinese seamen in cruise ship and tax preferential policy. In the aspect of research on cruise ship home port and industrial economy, Zhang (2008) discussed development mode of Shanghai's cruise ship home port, port layout, etc. Zhu (2013) focused on functional orientation of home port of Tianjin Port;

Zhu (2013) and Chen Jihong et al. (2012) focused on industrial development and bottleneck and functional construction of Shanghai Cruise Ship Home Port; Zeng et al. (2012) had a quantitative research on assessment index system for cruise ship home port construction. In the aspect of ship supply, Zhang (2016) and Zhu and Yan (2015) analyzed core competitive advantage for logistics supply chain of cruise industry and development and problems of Shanghai's cruise ship supply industry; Xu (2011) put forward policy demand in *Policy Appeal for Ship Supply Industry of Shanghai Port* according to *Port Law of the People's Republic of China* and No. [2009] 57 document of General Office of the State Council of the People's Republic of China; Wu (2009) analyzed shipbuilding structural condition from three aspects: industrial structure, industrial organization structure and product structure. In the aspect of supporting construction of cruise industry, Mai et al. (2009) estimated investment for building large and medium-sized internal cruise ship terminal and put forward measures for controlling investment risk; Wang et al. (2008) discussed planning and design of cruise ship home port. Zheng et al. (2006) analyzed current condition, problems and planning scheme of Shenzhen Shekou Passenger Terminal. Ma et al. (2009) calculated demand for cruise ship passenger capacity and construction scale of Shenzhen Shekou Passenger Terminal Area. Liu et al. (2007) described construction and feasibility of Tianjin Cruise Ship Home Port. Xu et al. (2008) researched silt digital analogy test of Tianjin Internal Cruise Ship Terminal Project in detail. Yu (2008) organized competitiveness and bottleneck for developing cruise ship home port in Dalian. Zhang (2016) organized features and bottleneck of China's cruise ship manufacturing industry.

2.2 Research Review of Cruise Supporting Industry

Through research on literatures about cruise supporting industry development at home and abroad, existing foreign research is mostly focused on development mode and experience reference of cruise industry, research on supporting industry in the whole industry chain is relatively little and relevant research on cruise supporting industry and service is made to some degree. However, in China, whether in cruise supporting industry or cruise supporting service, informative and systematical theoretical research and practical exploration are in shortage. Particularly for Shanghai Wusongkou International Cruise Terminal and other similar internal cruise ship home ports, research on cruise supporting industry and service is relatively lacking and required to be researched on a systematical, comprehensive and in-depth basis and forward-looking, scientific and strongly operable development path of cruise supporting industry has been formed.

3 Research Goal, Process and Method

3.1 Research Goal

Enrich research connotation and theoretical basis of cruise supporting industry; accelerate overall linking and upgrading of Shanghai's cruise ship industry chain; offer new research idea for quickening and improving Shanghai's overall capacity for blending into global cruise industrial value chain; provide reference for relevant policy and system formulation of cruise supporting industry.

3.2 Research Method

Based on member of research acting on general manager in Shanghai Wusongkou International Cruise Terminal and taking a temporary post in industrial planning office of Shanghai Municipal Tourism Administration and Baoshan Tourism Administration of Shanghai City, this paper takes Shanghai Wusongkou International Cruise Terminal as research object, seeks development bottleneck and gap of China's and Shanghai's cruise supporting industries by organizing famous relevant literatures for cruise supporting industry development at home and abroad; makes research interview and distributes questionnaires oriented to cruise supporting industrial construction and service, industrial park functional orientation and staged development emphases of main cruise ship port cities at home and abroad by yearly holding cruise industry conference and attending internal cruise industry conference; has a positive analysis for implementation path and key measures of cruise supporting industry; consults expert's suggestions and executes development research of Shanghai's cruise supporting industry based on research group members' going to the US, Canada, France, Japan, Hong Kong and other regions in China to investigate cruise ship terminals, cruise ship associations, cruise ship research institutions and universities and colleges and cruise ship companies and interviewing experts in cruise industry.

3.3 Research Process

From February 2016 to December 2016, distribute paper and electronic questioners. For paper questioners, distribute and recover them on the spot; parts of them were sent to organization investigated through express delivery and would be sent back after filled up. For electronic questionnaires, distribution mode was e-mail, web page link and WeChat QR code, quantity distributed was 320 and quantity of effective questionnaires recovered was 287. Average time consumption in survey and interview for each organization was more than 30 min. Interviews executed through site record, sound recording and other ways involved 76 persons in total and documentary data exceeded 30,000 words on an accumulative basis; data were analyzed after organized.

4 Research Conclusion

4.1 Development Features of Shanghai's Cruise Supporting Industry

4.1.1 China's Cruise Supporting Industry Chain Has a Rapid Development

Local cruise lines and self-owned cruise ship operation companies are preliminarily established. At the end of 2011, HNA Tourism Group became the first Chinese-funded company to purchase "Henna" from Carnival Corporation & PLC in the US and broke the situation for overseas cruise line monopolizing Chinese market; Ctrip cooperated with Royal Caribbean International to jointly establish "SkySea Holding International Ltd.". "Chinese Taishan" is the first wholly owned and independently operated and managed internal luxury cruise ship. In April 2016, China Communications Construction Company Limited, China National Travel Service (HK) Group Corporation and China Ocean Shipping (Group) Company jointly established Sanya Cruise Ship Development Company to create national brand of cruise industry. Cruise ship design, manufacturing and maintenance itch for a try. In September 2016, China State Shipbuilding Corporation, Carnival Corporation & PLC and Fincantieri jointly signed Letter of Intent for building 2 + 2 large luxury cruise ships, and the first large luxury cruise ship made in China is predicted to be put into operation in 2023. In 2015, China Merchants Group and Carnival Corporation & PLC signed memorandum of understanding to explore Chinese local cruise ship and its transformation; they fully seized investment opportunity that cruise ship manufacturing brought about and located cruise ship department in Shenzhen. In 2017, CMB Financial Leasing Co., Ltd., the subsidiary of China Merchants Group, closed a USD 500 million deal-building two cruise ships with Viking Cruises. Cruise ship supply, operation and sales are highly centralized. In terms of production value from the view of industry chain, cruise ship design and manufacturing, cruise ship operation and management and supporting service in terminal respectively account for 20, 50 and 30%. Upstream resources of China's cruise industry (except for port) are mostly dependent on foreign oligarch structure and cruise ship operators are highly centralized. Downstream travel agency (including OTA) plays a remarkable role and ship chartering and distribution become China's distinct cruise ship sales mode. In 2016, passenger volume of global cruise ships totaled 24.70 million and China became the second largest cruise ship passenger source market. China's online-cruise

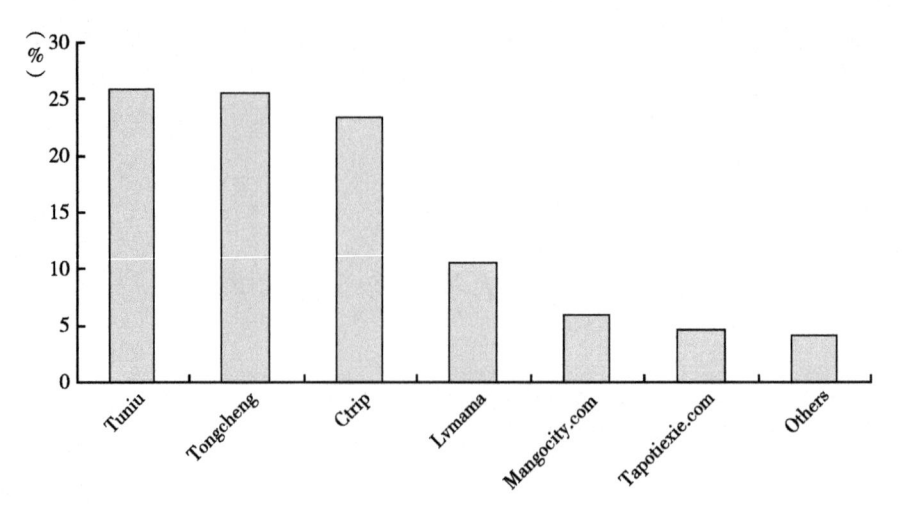

Fig. 1 Transaction scale proportion of online cruise ship distributor 2016H1 (%)

market gradually steps into rapid development stage and penetration rate of online cruise tourism will go on rising. In 2016, internet penetration rate of cruise market accounted for 60% or so, and in 2018, that number will be 65% according to prediction. Among China's current cruise ship distributors, tuniu. com, LY.com and ctrip.com perform well in existing resources, market coverage, operation capability, market expansion, etc. In the first half of 2016, market shares of the three companies accounted for 70% totally; market sharing of online distributors will get increasingly fierce, as shown in Fig. 1.

4.1.2 Agglomeration Effect for Shanghai's Cruise Supporting Industry Gradually Stands Out

In 2016, with total 1010 voyages of cruise ships, China became world's second largest cruise market; Shanghai, as the first experimental area for China's cruise tourism development, became the first largest cruise ship home port in Asian and the fourth largest one in the world, with status dramatically improved in global cruise industrial pattern, and developed into the critical stage for "transformation, upgrading and quality and efficiency improvement". It thus can be seen that agglomeration effect of cruise supporting industry gradually stands out. Optimization and upgrading of cruise supporting service software and hardware speed up. In September 2016, unveiling ceremony of CSSC International Cruise Industrial Park, the first internal cruise industrial park in China, was held in Shanghai. The industrial park is devoted to providing professional cruise ship manufacturing, operation and supporting service for China's cruise ship and in favor of optimization and upgrading of China's cruise ship full-industry chain. In May 2017, China State Shipbuilding Corporation, Carnival Corporation & PLC and

Baoshan District of Shanghai signed letter of intent for cooperation of cruise supporting industry to jointly promote development of cruise supporting industry system in Baoshan District, Shanghai, boost both parties' establishing cruise supporting joint venture enterprise in the park and support joint venture enterprises settled and registered in the park are provided with priority to blend into cruise supporting system developed through group cooperation. In January 2015, "Costa Atlantica" arrived in Shanghai for repairing, which realized Shanghai's first international cruise ship maintenance business and filled in the gap in cruise ship maintenance. In December 2015, Shanghai Huarun Dadong Shipyard Limited completed the initial overall modification for "MSC Geneva". In 2016, the Italian shipbuilding group Fincantieri signed exclusive cooperation agreement with Shanghai Huarun Dadong Shipyard Limited to provide service for maintenance and building of Chinese cruise ships. On May 29, 2016, VIP Service Center of Shanghai Wusongkou International Cruise Terminal was formally put into operation, which was aimed to offer full-process one-stop service to tourists; moreover, Linjiang Boutique Hotel and Linjiang Hotel particularly providing supporting service for Shanghai Wusongkou International Cruise Terminal were also put into operation. In September 2016, Shanghai International Cruise Tourism Service Center was unveiled and formally put into operation, a center devoted to building online and offline integrated smart cruise tourism service platform which will be the most authoritative and has the greatest cruise ship superior resources in China. Cruise supporting policy and ship supply service offer the great convenience. In 2016, according to approval of State Council, cruise ship entry visa policy got looser after implementation of "144 h Visa-free Transit for Foreigners" and "15-day Visa Exemption for Foreign Tour Group Entering China by Cruise Ship". In 2017, Shanghai Cruise Port Railway Internal Tourism (Group) Co., Ltd. cooperated with Shanghai International Cruise Tourism Service Center Co., Ltd. to jointly open HSR-cruise ship through train and offer "aviation-railway-port" integrated service. Meanwhile, Shanghai continues accelerating base construction of cruise ship headquarters, constantly strengthens investment promotion and successfully introduces cruise ship operation enterprise owned by Ctrip; registers and establishes SkySea Holding International Ltd. in Baoshan District and launches currently largest Chinese cruise ship booking platform in the world. Finally, Shanghai Port has number of enterprises involved in port ship supply more than 400 and becomes the port having the most number of port ship supply enterprises, largest operation scale and most number of practitioners in China. In 2015, Shanghai's first cruise ship food transit supply was implemented in Shanghai Wusongkou International Cruise Terminal, which was the first time that Shanghai supplies transit food to international cruise ship. Additionally, Wusongkou International Cruise Terminal Development Co., Ltd. and Costa Cruise Lines cooperate to have a successful try for container transshipment business of Royal Caribbean Cruises Ltd., have cruise ship supply, warehousing enterprise, distribution center and other businesses rapidly developed and are strongly supported by convenient customs clearance service, cruise ship supply and materials distribution procedures, transit inspection and quarantine and other relevant departments.

They focus on cultivation and improvement of cruise supply logistics and intensively improve level of cruise ship supply logistics; at that time, they will develop from the stage for ship supplying parts of fresh food and consumables into the important system for global cruise ship supply. Cruise supporting financial service provides the strong support. In December 2016, China State Shipbuilding Corporation led and coordinated six parties to jointly launch and establish 30 billion yuan China's first cruise industrial fund and register it in Baoshan, Shanghai, which provided the strong support for Shanghai's cruise supporting industry's actively docking development strategy of *Made in China 2025* and accelerating transformation and upgrading of Shanghai's supporting industry.

4.2 Bottleneck for Shanghai's Cruise Supporting Industry Development

4.2.1 Unification and Coordination Mechanism for Port Operation Is Required to Be Improved

It is unable to be unified and coordinated in port operation and management. Operation and management of cruise ship terminals is involved with sanitary inspection and quarantine, customs, frontier inspection, public security, tourism, maritime affairs and other departments as well as travel agencies, cruise lines, commercial companies, labor service companies and other commercial entities. Competent authority is diversified, management standard and specification are hard to be unified and it is difficult to ensure the whole operation process is scientific and effective in case of relying on port only. Particularly under unexpected situation due to high passenger flow peak and force majeure, cruise ship cannot provide a timely, effective and united information exchange platform or a uniform and coordinated management mechanism. According to survey and review, breakthrough should be made in port customs clearance facilitation. At present, 144 h visa-free transit policy is implemented in Shanghai's cruise tourism. According to the policy, foreign tourist entering China cannot be allowed to enjoy relevant visa-free transit policy unless he or she is going to the third country, and it, objectively, brings about lots of inconveniences to foreign tourists in voyage arrangement. Effect of visa-free transit policy is required to be enlarged and breakthrough should be made in cruise ship direct visa exemption policy of visiting port; entry procedure is required to be further facilitated and both convenience and safety should be considered. System innovation, copy and promotion of Shanghai Free Trade Zone are required to be strengthened. System innovation of Shanghai Free Trade Zone is involved with expansion and opening up of six fields in service industry, system innovation for investment and management, goods trade facilitation and other measures; but opening up, goods trade facilitation and other measurers of professional service industry closely related to cruise industry development cannot be copied and spread to China's cruise tourism development experimental area. Cruise ship supply market and ship supply supervision mode are required to be established. Main conflict for Shanghai's cruise ship supply market is mainly in great difference between supply and demand of goods supplied by cruise ship in quality, quantity, variety and service level. Since tax reimbursement cannot be enjoyed in local purchase of cruise ship supply, enterprise's purchase cost is increased, and industrial monopoly exists objectively. In addition, construction of international cruise ship materials distribution center is the principal component for cruise ship economy development. Currently, world's large cruise lines have distribution mode focused on centralized procurement and directional goods distribution. There is no relevant documents and operating methods particularly oriented to cruise ship supply trade facilitation in current China, and existing laws and regulations are lacking of policy support for distribution mode above; consequently, entry food supply of international cruise ship cannot be realized in Shanghai.

4.2.2 Clustering Degree of Industrial Factor Is Low and Regional Economic Linkage Is Required to Be Strengthened

Domestic cruise ship economy has a late start in development and clustering degree of cruise ship economic factor is low as a whole; clustering scale effect of cruise industrial factor has not been formed and radiation and pulling function for surrounding area is required to be strengthened. Cruise supporting industrial park and city-industry integration are required to be improved. (1) Problem about surrounding land using of cruise ship terminal: Land ownership is complex and lots of efforts should be made in land exploitation. (2) Problem about surrounding traffic layout and commercial layout of cruise ship terminal: The new normal for simultaneously pulling into shore of two and three ships and even four ships in the future sharply increases the traffic pressure for port connecting the trunk road Baoyang Road, traffic jam at ship berthing rush hours cannot be solved and there is no aerial train with sightseeing function at present; besides, there is a small number of surrounding commercial facilities of cruise ship terminal and united commercial layout planning is lacking. (3) Problem about cruise ship terminal investment and operation: Foreign cruise ship terminal is a kind of public facility and mainly funded by nation and regional government, e.g. Hong Kong invests HKD 8 billion to establish new cruise ship terminal Qingdao Port, Tianjin Port and Xiamen Port are funded by municipal government. Investment for Phase I and II of Shanghai Wusongkou International Cruise Terminal are mainly from district-level financial investment and future development will be influenced.

4.2.3 The Space for Cruise Ship Maintenance, Manufacturing and Modification Working on Cruise Ship Building Localization Is Vast

Under the background that China accelerates cruise tourism development, localization trend of China's cruise industry is obvious and localization trend of cruise supporting industry is irresistible. China's cruise ship repair, modification and manufacturing, as "the last kilometer" for ship repair industry's transforming and developing towards high-end industry, does not become "new normal" of China's ship repair enterprise, and ship repair enterprise has a large space in cruise ship maintenance technology, modification, manufacturing, indoor decoration and design and other matching fields. Meanwhile, according to survey and interview, in 2015, Shanghai Huarun Dadong Shipyard Limited, when repairing the cruise ship "Costa Atlantica" and completing overall modification for the cruise ship "MSC Geneva", attract professional technicians of design institutes and enterprises for visiting and learning. COSCO and China Shipping Industry Co., Ltd. and other backbone enterprise repair enterprises adopt "repairing and manufacturing integrated" development strategy, and it is possible that cruise ship repair, manufacturing and modification experience must accelerate and promote cruise ship building localization.

4.2.4 Jam in Yangtze Estuary Channel Is Increasingly Obvious and Navigation Capacity Is Faced with the Great Challenge

Currently, Yangtze Estuary Channel is challenged with high navigation density, diversified ship variety and constantly improved management difficulty. The main channel for ship passing through Yangtze Estuary-Beicao Channel has navigation capacity tending to be saturated; where ship width is more than 40 m, it will be hard to arrange two ships passing through Beicao Channel simultaneously and navigation resources are extremely limited. With the trend for large cruise ship being obvious in the world and quantity of cruise ships berthed in Shanghai Wusongkou International Cruise Terminal rapidly increasing, annual number of cruise ships berthed in Shanghai Wusongkou International Cruise Terminal will be 600 in 2020 according to prediction, navigation ability of Yangtze Estuary will be face with a great challenge. Therefore, the exploration for enlarged channel, schemed channel management and normalized emergency treatment mechanism at Yangtze Estuary will get increasingly urgent.

5 Discussion and Analysis

5.1 Optimization Path for Shanghai's Cruise Supporting Industry

5.1.1 Optimization Path for Cruise Supporting Industrial Park

Gradually advance park area and urban area integrated development path based on CSSC International Cruise Industrial Park and "port-city-area"; focus on cruise ship full industry chain, strengthen independent development in upstream link of industry chain, cultivate headquarters economy in core link of industry chain, accelerate supporting development in operation link of cruise ship home port and improve function in cruise tourism operation link; establish modern service system of cruise ship, innovate cruise ship investment and management system and supervision system, advance "zone-terminal linkage" development and establish cruise industrial system focused on cruise ship service and modern service ecological cluster. Strengthen international cooperation and inter-industry cooperation based on "the Belt and Road" initiative.

5.1.2 Layout Optimization of Cruise Industrial Supporting Space

For cruise supporting industry, fully consider "water and shore linkage and port and city integration" on the basis of completely referring to Urban-rural Overall Planning Outline of Baoshan District, Shanghai (2015–2040); adjust Baoyang Road Chongming Wharf into passenger transportation center of cruise ship terminal; create integrated transportation hub of Shanghai Wusongkou International Cruise Terminal based on Baoyang Road integrated transportation hub; establish five areas in space layout, i.e. cruise ship home port core area, northern cruise supporting service area, southern cruise supporting service area, cruise ship and yacht materials distribution center, cruise ship and yacht training base and design R&D center. Accelerate cruise industrial adjustment and create riverside service industry development belt characterized with cruise tourism; extend cruise industry chain and establish Shanghai cruise tourism cultural area characterized with local cruise ship operation and cruise tourism with Chinese characteristics.

5.2 Optimization Countermeasures and Suggestions of Shanghai Cruise Supporting Industry

5.2.1 "Zone-Terminal Linkage" System Innovation Accelerates Structural Optimization of Cruise Supporting Industry

Baoshan district of Shanghai should continue deepening and innovating "zone-terminal linkage" and further collect factors of cruise ship economy; formulate relevant preferential policies and measures and advance financial innovation system in favor of settlement and development of cruise ship company headquarters. Meanwhile, it is required to actively establish "global purchase-local transshipment" supply chain, lower internal cruise ship materials purchase cost and establish global cruise ship materials supply system; further improve discourse power in global cruise ship materials supply and establish Shanghai cruise ship materials distribution center on the basis of current ship supply service; expand internal cruise ship company's purchasing scale in China, optimize structure of cruise supporting industry and add more vitality to regional economic development.

5.2.2 Optimization and Upgrading of Cruise Supporting Service Industry

Cultivate cruise ship service industry chain vigorously. Accelerate standard construction of cruise tourism and development of specialized cruise travel agencies; improve port service level of cruise ship terminal, normalize order of cruise tourism market and strengthen tourism safety assurance system construction. Try to expand cruise ship service industry chain. Collect high-end resources and factors of cruise industry, start construction of internal cruise ship terminal and large theme park as soon as possible; further increase quantity of yacht wharfs and expand coverage of cruise ship route. Meanwhile, accelerate installation of various relevant facilities for cruise ship supply, waste disposal, maintenance and repair and establishment of cruise ship base and cruise line; establish corresponding transportation and tourism service facilities and try to create development and operation environment of cruise industry conforming to international practices. Strengthen cruise ship operation industry chain. Actively introduce and establish cruise lines and welcome more cruise lines and industrial associations to settle in Shanghai; encourage and support companies in cruise industry chain to cooperate and execute cruise ship operation; offer interest, tax, finance, credit and other preferential supports in ship chartering and purchase.

5.2.3 Optimization of Cruise Supporting Facility and Service

Relevant supporting service optimization for core function of cruise industry: Strengthen cruise ship infrastructures construction, improve cruise ship terminal quality and attract settlement of internal shipping enterprise; reasonably plan and develop buildings and supporting facilities in cruise ship terminal area and surrounding hinterland and realize supporting facility and service optimization of cruise ship political and commercial information platform. Optimization of supporting facility and service driven by cruise industry: Optimize cruise ship education and training as well as supporting facility and service of research center; accelerate construction of public service system and emergency system, actively promote construction of cruise ship distribution center and optimization of supporting facility and service of cruise ship finance and insurance support center and create cruise ship finance and insurance center in line of Shanghai World Financial Center.

5.2.4 Energy Level Improvement of Cruise Supporting Industry

Actively absorb cruise line's and foreign cruise ship terminal's international capital, domestic commercial capital, investment fund, individual investment, etc.; accelerate cruise ship terminal finance and listing, construct cruise ship follow-up project according to high standard and improve operation capacity of cruise ship terminal and reception capacity and service level of Shanghai Wusongkou International Cruise Terminal. Accelerate improvement of cruise ship home port function and promote factor agglomeration of cruise industry; establish cruise ship ticket center, materials supply center and talent service center in Shanghai. Accelerate and promote construction of international cruise ship city and overall transformation of riverside development belt; focus on key projects, establish Long Beach of Shanghai and accelerate and promote construction of cruise ship.

5.3 Key Measures for Optimization of Shanghai's Supporting Industry

5.3.1 Establish Shanghai Wusong International Cruise Ship Supply Bonded Logistics Area

Establish and operate cruise ship supply system overall on cruise ship service supply chain and improve customers, national inspection, frontier inspection, maritime affairs, tax, business and other government management functions related to cruise ship and ship supply supervision; focus on finance, settlement, insurance and other supporting commercial functions and services; seize the opportunity for constructing "Experimental Area for China's Cruise Tourism Development" to establish Shanghai Wusong International Cruise Ship Supply Bonded Logistics Area.

5.3.2 Establish China Cruise Tourism Cross-Border Commodities Exchange Center and E-Commerce Platform

Strengthen cooperation with main cruise ship terminals in Asia; introduce "internet+" idea by docking relevant policies of free trade zone and duty-free shop; explore and establish Shanghai cruise tourism cross-border commodities exchange center and e-commerce platform; optimize payment function system and realize transformation from traditional retail business to "experiential marketing" and efficient integration of online and offline tourism commodities; innovate industrial clustering platform, improve shopping service experience of cruise tourists and give full play to cruise ship home port's pulling function on regional economy and radiation effect on relevant industries.

5.3.3 Establish Cruise Ship Service Talent Supply Base

Actively cultivate skilled, service and management talents required in development of Shanghai's cruise industry, formulate relevant policies to attract excellent talents at home and abroad and provide support and power for sustainable development of Shanghai's cruise tourism; improve service level and market competitiveness of Shanghai's cruise ship terminal; attract senior operation, management and technology R&D talents of foreign cruise line to come to Shanghai for work; strengthen school-enterprise cooperation and promote universities and colleges and cruise lines to jointly establish cruise ship training base and offer training, internship and research opportunities; establish national base for cruise ship operation, management, R&D and manufacturing and cruise crew training; promote cruise ship terminal's cooperating with world's famous cruise line, cruise ship terminal and travel agency and improve vocational quality of existing cruise ship talents; establish cruise ship service talent supply base to improve service level and market competitiveness of cruise ship terminal.

5.3.4 Establish Shanghai Cruise Ship Repair and Building Supporting Base

As indicated by development experience of European and American countries, cruise ship repair and building, belonging to high-end tourism equipment manufacturing industry, can yield considerable earnings. Establish Shanghai cruise ship repair and building supporting base and actively participate in development alliance of China's cruise ship manufacturing industry; issue preferential policy to attract cruise ship design and building enterprises to settle in Baoshan District and improve value chain of cruise industry. Establish think tank of cruise ship repair and building supporting base and actively invite senior executives of cruise line, cruise ship manufacturing industry, famous enterprise for cruise ship design, cruise supporting industry and classification society to be involved in construction of base think tank; discuss development trend and latest technology for global cruise ship building and manufacturing, repair and building supporting facility and supply system and establish think tank for cruise ship repair and building supporting base. Establish China's cruise ship repair and building and accessories supply center Establish China's cruise ship repair and building and accessories supply center based on the trend for world's cruise ship repair and building industry moving eastwards, Shanghai's long-standing shipbuilding advantage and Baoshan District's excellent industrial foundation, so as to provide accessories supply and repair and building services for cruise ship repair and building in China and even Asian-Pacific region. Accelerate optimization and upgrading of cruise supporting industry chain and establishment of China's discourse power and initiative in global cruise ship repair and building industry.

Chapter 9
Research on the Financial Service System for Development of China's Cruise Industry

Yingkai Yin, Zheng Yu and Zhihui Jiang

Preface

Modern cruise industry has become one of the most high-end business with the fastest growing speed and the most potential in the international tourism market. Because of its strong driving and adsorption capacity, the cruise industry has become an important growth point of the city economy and an important new energy for the economic growth of one country. Thus, it is known as "the Gold Industry floating along the Golden Waterway".

China's cruise industry started since 1990s and entered the stage of vigorous development in 2006. Now it has become one the fastest growing points in the world. In 2016, the total number of tourists on the global cruise market reached to 24.7 million, with a year-on-year increase of 6.5%. China's annual cruise tourists reached to 2.1 million, becoming the second largest cruise market in the world.

At the same time, China's cruise industry still has chock points. For example, the industry chain is not completed and it is in the low end of the global cruise value chain. Finance is the core of the modern economy. The cruise industry is capital and technology-intensive. Thus, the sound financial service system is very important to the development of the cruise industry and is also conducive to break the above-mentioned chock points.

Yingkai Yin—Research field: International finance, Green finance and financial technology.
Zheng Yu—Research field: Digital inclusive finance.
Zhihui Jiang—Research field: Green finance.

Y. Yin (✉) · Z. Yu · Z. Jiang
School of Economics, Shanghai University, Shanghai, China
e-mail: yinyk@i.shu.edu.cn

1 Literature Review

1.1 Research at Home and Abroad

1.1.1 The Overall Development of the Cruise Industry

Compared to the rapid development of the cruise industry in recent decades, its research is lagging behind. In 2011, there were two summarized papers on the cruise tourism. Papathanassis and Beckmann (2011) reviewed the literature of the cruise tourism from the perspective of tourism research and put forward the research dilemma and the feasible research process of the cruise tourism. Sun et al. (2011) reviewed the current situation of the cruise industry from the perspective of market research, cruise operation and revenue management.

In China, Sun and Feng (2012) reviewed the research problems and achievements of the cruise tourism industry in China based on the existing domestic literature. Since 2009, China Cruise and Yacht Industry Association (CCYIA) and Shanghai International Shipping Institute have jointly issued *Annual Report of China's Cruise Development* for seven consecutive years. Since 2014, Shanghai International Cruise Business Institute has begun to prepare *China Cruise Industry Development Report* each year, showing the overall development of China's cruise economy in a whole year from the perspective of regional development of cruise, international experience and cruise industry chain (Wang et al. 2014, 2015, 2016). These studies believe that in recent years, China's coastal areas, seizing the opportunity that the world cruise center quickly moving to the east, accelerated improvement and construction of the cruise port city, promoted and enriched the extension development of the cruise industry chain and gradually formed and improved the supporting system of the cruise industry, which strongly support the long-term sustainable development of China's cruise industry. With the transformation and upgrading of China's economy, the potential of cruise tourism market will continue to be released in the future. However, the chock points of the domestic cruise manufacturing needs to be settled. Emphasis should be placed on the introduction of international capital-related policies and legislation.

1.1.2 The Demand of the Cruise Industry for financial Support

The financial needs of the cruise industry mainly include investment and financing, insurance and convenient transaction (such as remittance settlement and bank card business). Zhao (2011) studied the innovative ship financing way and expounded the problem of China's taxation policy in financing and rental of the ship. Xiao (2014) discussed each characteristic the risk assessment system should have. Nevitt and Fabozzi (2000) pointed out that, in western countries, the cruise financing industry is relatively mature. There are investment banks with the core business of providing ship financing service, specialized cruise industry fund and sound policy support in these countries. Furthermore, the investment bank has provided sufficient funds for cruise financing.

1.1.3 Financial Service System of China's Cruise Industry

In recent years, with the rapid development of China's cruise industry, some literatures conducted a partial study on the financial services system for the development of the cruise industry. Wang et al. (2014) believed that the corresponding financial and credit support policies should be formulated. Firstly, establish the sound special cruise directory in the field of insurance, credit and fund. Secondly, set up the single-ship experiment financing and leasing company and agree cruise lines registered in China to set up offshore accounts. Thirdly, solve the financing problems through the bank financing, cruise industry funds or cruise trust products to support innovation of financial products of the cruise. Sun (2014) believed that there are financial credit barrier and risk management barrier in China. Huang and Zhu (2014) thought that Shanghai needs to focus on the cruise tourism in the special tourism development fund and to set up the cruise industry fund and gradually expand if it wants to develop the cruise industry. Jiang (2014) introduced the "balance held on local currency and foreign currency deposit in financial Institutions" and "the insured amount in insurance agencies" as the important indicator to evaluate the competitiveness of China's cruise port market.

1.2 Research Review

We can see from the above review that on the one hand with the rapid development of the cruise industry, research on the cruise industry is increasing; on the other hand, some literatures have begun to pay attention to the financial service system of the cruise industry. These literatures provide a good foundation for this paper.

However, in general, compared with the rapid development of the cruise industry, the current research is still insufficient and lagging behind, and the study on financial service system is still immethodical. In this paper, based on the perspective of extending the value chain of China's cruise industry and improving the position of China's cruise industry in the global value chain, the core question "financial service system for the development of China's cruise industry" is focused on, systematic sorting is conducted to connotation, mode and path of the financial service system of the cruise industry, and policy suggestions are provided to the construction of financial service system of the cruise industry of China, especially Shanghai.

2 The Present Situation and Shortcomings of China's Cruise Industry

2.1 Cruise Economy Is Developing Rapidly in China and Shanghai

In recent years, China's cruise industry was booming and has become one of the fastest growth points in the world, with increasing position in the world cruise economic pattern. In 2016, the number of cruise ship China's top ten cruise ports received was 996 in total (with a year-on-year increase of 58%) and the number of cruise tourists was 2.26 million (with a year-on-year increase of 82%). China's cruise market is currently close to 50% of the total Asian market. The international market share increased from 0.5% in 2006 to 0.9% in 2016, which has increased by 20 times against ten years ago, ranking eighth in the whole world. Over the past decade, the number of Chinese who chose to travel through the cruise has increased 320 times and the average annual compound growth rate exceeded 40%. It is expected that by 2030, the number of Chinese who choose to travel by cruise ship will be close to 20 million, and China is expected to become the world's largest cruise tourism market.

Shanghai is the center of China's offshore cruise routes and Asian cruise routes and also the important bellwether of "the Belt and Road". It has huge and growing consumer market in the Yangtze River Delta, complete three-dimensional transport network, historical background and modern elements. In recent years, the Shanghai cruise market has developed rapidly. In 2014, Shanghai Home Port exceeded New York, becoming one of the world's top eight world-class home ports of cruise ships. In 2015, the number of international cruise ship entering and leaving China (port) was 688 ships, with a year-on-year increase of 26.9%. In 2016, in China's top 10 ports, Shanghai's total voyages accounted for 51% of that of the whole country and its Chinese and foreign tourists accounted for 65% of the country. In 2016, Shanghai Baoshan Wusongkou International Cruise Port received 471 cruise ships in total, accounting for 47% of that of China and the number of inbound and outbound tourists it received was 4.515 million, accounting for 63% of that of China, with a year-on-year increase of 69 and 87% respectively, remaining the first in Asia.

2.2 Serious Shortage Behind the Rapid Development

China's cruise industry obtained its strategic development period under the driving of demands, the guidance of government and the support of policy. It is expected that in the future, the cruise industry will drive the economic growth scale of over 100 billion yuan. There are still two prominent problems behind the rapid development. Firstly, the complete cruise industry chain has not been formed and the pulling effect on the economy is not obvious. In China, most cruise ports are used for "cruise tourism". Secondly, China's cruise industry is still at the low end of the value chain of the global cruise industry. Its main business is cruise ship berthing and tourist reception, which are the downstream of the industry chain. It just started in the upstream link such as design, manufacturing and decoration of the high additional value cruise ship. Thus, it is an important strategic issue for the development of the cruise industry in China to build the wide industry chain, effectively embed the global value chain and to climb the high value-added link and play the pulling effect of the cruise industry.

2.2.1 Connotation of the Whole Industry Chain

At present, with the rapid development of the global cruise industry, the cruise industry chain continues to expand to the upstream, midstream and downstream. The upstream refers to design and manufacturing of the cruise ship and the upstream enterprise is mainly cruise ship manufacturer. The midstream is construction of the cruise port and cruise ship operations and travel services provided by cruise operation companies, and the midstream enterprise is mainly the port terminal and the cruise operation company. The downstream mainly refers to cruise tourists reception and other auxiliary services, including organization and transportation of cruise ship passengers and reception services at the port terminal, and the downstream enterprise is mainly the travel agency, direct sales team of the cruise ship operation company, etc.

In the value chain competition of the global cruise industry, the appreciation range of the upstream and midstream continues to show. For example, the output value of the cruise design and manufacturing link accounts for about 20% of the entire cruise industry chain and the market demand is broad (Fig. 1).

2.2.2 Current Situation of the Whole Cruise Industry Chain in China

In 2006, the first international modern cruise ship docked in China. Since then, the domestic cruise tourism market grew rapidly at an average annual rate of about 40%. China's cruise industry experienced the budding stage, the tourism blowout stage and the extension and expansion stage in the ten years of rapid development.

Design and building of the cruise shipDesign and building of cruise ship	Construction of home port & operation of cruise shipConstruction of home port and operation of cruise ship	Cruise tourism reception serviceCruise tourism reception service
Upstream	Midstream	Downstream

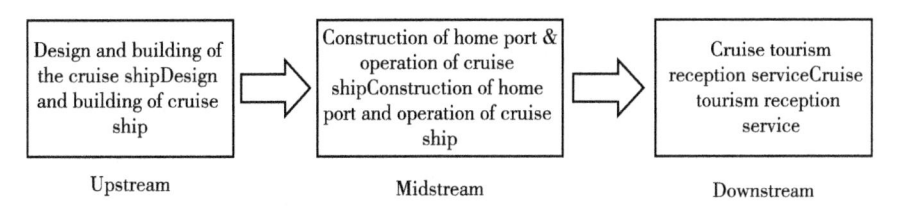

Fig. 1 Whole industry chain of the cruise industry

In the next decade, China is expected to exceed the United States, becoming the world's largest cruise tourism market.

On the one hand, China's cruise industry has a lot of room for development. In 2016, China's per capita GDP reached to USD 8126. In some coastal provinces and cities, the per capita GDP exceeded USD 10,000. With these conditions, the cruise tourism can develop rapidly. However, according to data of the Cruise Lines International Association, at present, the penetration rate of the cruise tourism market in North America and Europe is 3.2 and 2% respectively, and 3.5 and 3.4% respectively in America and Australia. However, the penetration rate in China's cruise tourism market is less than 0.05%, which is expected to maintain rapid growth momentum in the future.

However, on the other hand, the business China's cruise industry is mainly at the low end of the value chain of the global cruise industry. China is mainly responsible for undertaking foreign cruise lines and port services, which account for not large in the inbound and outbound travel. It is still weak in the upstream and midstream with high value and high additional value, especially design, manufacturing and decoration of the cruise ship. China's own cruise industry has just started. The Chinese-owned cruise operation company has just started in 2013 and the market share was less than 5% in 2015.

2.2.3 Current Situation of the Whole Cruise Industry Chain in Shanghai

Shanghai is the place where the largest home port for the cruise ship is located in China, even Asia. About 70% of the tourists traveling by the cruise ship every year start from Shanghai. Wusongkou International Cruise Port, in Baoshan District of Shanghai, has stood in the "spire" of the cruise industry of China, even Asia, becoming the most busy international cruise port in the Asia-Pacific region. Nowadays, the cruise industry of Wusongkou International Cruise Port has gradually become the new engine of Baoshan and Shanghai's economic growth.

Upstream of the Industry Chain

Shanghai is still weak in design and manufacturing of the cruise ship, which is its "short board".

It is actively "filling the short board" in this aspect. In 2016, CSSC, CIC and Carnival Cruise Group signed an agreement to develop China's domestic luxury cruise brand; CSSC and Fincantieri started to cooperate in the manufacturing technology; Shanghai Waigaoqiao Shipbuilding Co., Ltd. prepared to manufacture China's first luxury cruise ship relying on its advantages on shipbuilding. In 2016, Shanghai CSSC International Cruise Industrial Park officially opened to the public, which means that Baoshan District of Shanghai further extended its cruise industry chain to the cruise ship manufacturing and operation and so on.

Midstream of the Industry Chain

In the midstream of the cruise industry, Wusongkou International Cruise Port is mainly responsible for international cruise ship berthing and tourist reception and the throughput capacity increases year by year. Its industry chain is not long enough and the range is not wide enough. Cruise port construction and cruise operation are the core links of the industry chain of the cruise. In this regard, the most prominent problem of Shanghai is that the cruise ship in the home port and the cruise ship company are not large enough. At present, the world's three major cruise groups have set up offices in Shanghai, but only a few Sino-Japanese joint venture cruise lines regard it as the home port.

Thus, it is important to expand the whole industry chain in Shanghai and develop the cruise economy cluster. In the midstream of the industry chain, it is important to introduce the headquarters of international cruise lines to attract the cruise line to participate in construction of Baoshan cruise headquarters base, so as to give full play to the position advantage of the home port. When the home port has enough cruise line headquarters settled, the headquarters economy will drive the tourism service industry, which is in the downstream of the industry chain, to develop and grow, achieving the effect of agglomeration economy.

Downstream of the Industry Chain

The downstream of the industry chain will directly determine the choice of tourists to the cruise port because it is the link the port directly contacts with tourists. In recent years, the number of inbound and outbound tourists received by Wusongkou International Cruise Port has increased year by year, but the construction of supporting facilities in the port is still to be strengthened, including the tourist management system, standardization of tourism services, e-commerce platform construction, construction of cross-border commodity trading center and cruise travel international business level.

2.3 *Financial Service System of China's Cruise Industry Obviously Lagging Behind*

Finance is the core of the modern economy. The cruise industry is capital and technology-intensive. Thus, the sound financial service system is critical to the development of the cruise industry. China's cruise financial service has shortcomings such as "fragmented system and insufficient depth and support efforts".

2.3.1 Research on the Connotation of Financial Service System for the Development of the Cruise Industry

The connotation of financial service system for the development of the cruise industry can be analyzed from two dimensions: first is the upstream, midstream and downstream of the whole industry chain and second is investment and financing service, risk management service and convenient transaction of the financial service. Based on the above two dimensions, we can use the "nine-grid pattern" shown in Fig. 2 to describe the connotation of the cruise financial service system.

The first is design and manufacturing of the cruise ship. Design and manufacturing of the cruise ship are capital and technology-intensive. The capital investment is extremely large, the requirement to technology is high, the manufacturing period is long and the recovery of cost is relatively slow. Thus, design and manufacturing of the cruise ship needs to be supported by financial service, including investment and financing service and risk management service.

The second is the cruise ship operation and construction of home port in the midstream, including maritime operations, capital operations and operation and management of the offshore project. The annual operating cost of international

Financial service / Value range	Investment and financing service	Risk management service	Convenient transaction
Upstream: Design and building of the cruise ship	Venture capital, syndicated loans, export credits, finance leases, corporate listing and revolving credit loans	The risk compensation mechanism for the first set of major technical equipment, cruise ship building insurance and cruise terminal liability insurance	—
Midstream: Construction of home port and operation of the cruise ship	Capital market financing (issuance of corporate bonds, corporate bonds, debt financing instruments, project income bonds, corporate listing), financial leasing, government financial investment and special funds	Cruise ship shipping insurance, cruise port infrastructure insurance and pollution liability insurance	—
Downstream: Cruise tourism reception service	Consumer credit and other consumer finance	Cruise ship cross-border theft, cruise carrier passenger liability, tourist insurance and other cruise insurances, (in the event of typhoons and other emergencies) route risk and group risk	Exchange settlement, bank card service, payment and deposit

Fig. 2 "Nine-grid pattern" for connotation of financial service system for development of the cruise industry

cruise lines is basically at around USD 13–14 billion. To build a large and medium-sized cruise terminal, about 1.5–2.0 billion yuan should be invested. In the area of cruise operation and home port construction, investment and financing services include capital market financing, central and local financial investment, the establishment of special funds, financial leasing, etc. Risk management services mainly involve multi-class cruise operations insurance.

The third is the cruise tourism reception service in the downstream, including a series of reception, catering, accommodation, sightseeing and shopping from the beginning to the end. Financial services in this area are mainly related to visitors, including the demand of tourists for exchange settlement, bank cards, credit cards and other convenient transaction and lending financial services. At the same time, the insurance service in the carriage process is also taken seriously. The development of Internet finance will provide cruise tourism with more convenient financial services.

2.3.2 Defects in China's Cruise Financial Service System: Based on the Comparison of Chinese and Foreign Enterprises

In general, China's cruise financial service has shortcomings such as "fragmented system and insufficient depth and support efforts".

In this paper, in order to further study the shortcomings of China's cruise financial service system, three domestic listed companies are selected from the upstream, midstream and downstream of the industry chain (CSSC, Bohai Ferry and Caissa Touristic) and three international counterpart enterprises (AKPS, CCL and TUI) are selected for comparison (Table 1).

The Upstream of the Industry Chain (Cruise Ship Manufacturing):
CSSC and Aker

Contrastive Analysis of Key financing Modes

The main financing mode of CSSC[1] is concentrated on all types of bank loans (80%) (including mortgages and pledged loans), government subsidies (accounting for 10%, including product subsidies, import discounts, R&D funding subsidies and infrastructure projects support, etc.). The proportion of other financing methods is relatively small (Figs. 3 and 4).

Aker,[2] the largest cruise ship manufacturer in Europe, is an international benchmarking enterprise. It has many financing channels, which not only includes secondary market financing (20%) and bonds (40%), but also includes financial leasing, revolving financing loans and policy loans. The developed European bond

---

[1]According to the summary of CSSC's annual report, the percentage is estimated.

[2]According to the summary of Aker Solution's annual report, the percentage is estimated.

Table 1 Basic information of case enterprise of the upstream, midstream and downstream of the cruise industry

Upstream of the industry chain	Design and manufacturing of the cruise ship	
Company name	CSSC (stock code: 600150)	Aker Solution (Oslo Stock Exchanges: AKPS)
Introduction to the company	In 2016, CSSC signed an agreement with Italy Fincantieri Group to establish a luxury cruise ship design and manufacturing joint venture company, investing and establishing "CSSC Cruise Industry Development Co., Ltd." in Shanghai	Aker Yards under the Group is the largest cruise ship factory in Europe and one of the world's five largest shipyards at the same time, mainly undertaking the construction of the ship with complex process and the most commercial value
Midstream of the industry chain	Operation of the cruise ship	
Company name	Bohai Ferry (stock code: 603167)	Carnival Corporation & PLC (The New York Stock Exchange: CCL)
Introduction to the company	In 2013, Bohai Ferry purchased "Traveler" from Costa and changed it to "Chinese Taishan". This ship had its maiden trip from Yantai in August of the same year	American Carnival Group is the world's first super luxury cruise line. It now has 24 large luxury cruise ships of 80,000–120,000 t and is called "The King of Cruise Ships"
Downstream of the industry chain	Cruise tourism reception service	
Company name	Caissa Touristic (stock code: 000796)	Touristik Union International (Frankfurt Stock Exchange TUI)
Introduction to the company	Caissa Touristic, a listed company under HNA, set up the cruise sales company, covering Tianjin, Shanghai, Hong Kong, Xiamen and other ports, providing integrated cruise service	TUI is Europe's largest tourism group and one of the world's largest tourism service enterprises. It has six luxury ocean-going cruise ships, providing professional and quality all-round tourism services

market has facilitated the financing of cruise shipbuilding enterprises. At the same time, European governments and national development banks provide low-interest loans and financing support for cruise construction projects to help cruise lines solve their "financing difficulties".

Breakthrough Path

From the experience of developed countries, foreign cruise manufacturing enterprises have good financial environment of "policy finance and market finance go hand in hand", rich financing channels and enough money. The financial environment of China's design and manufacturing enterprise still has limitations:

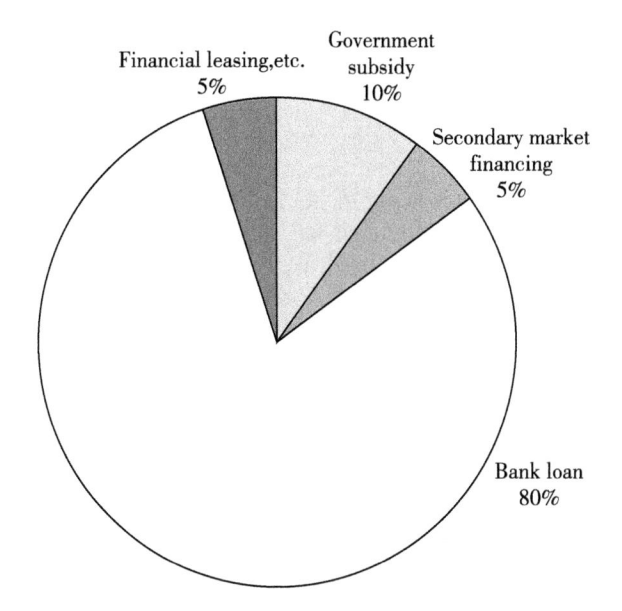

9 Research on the Financial Service System for Development ...

Fig. 3 Main financing structure of CSSC

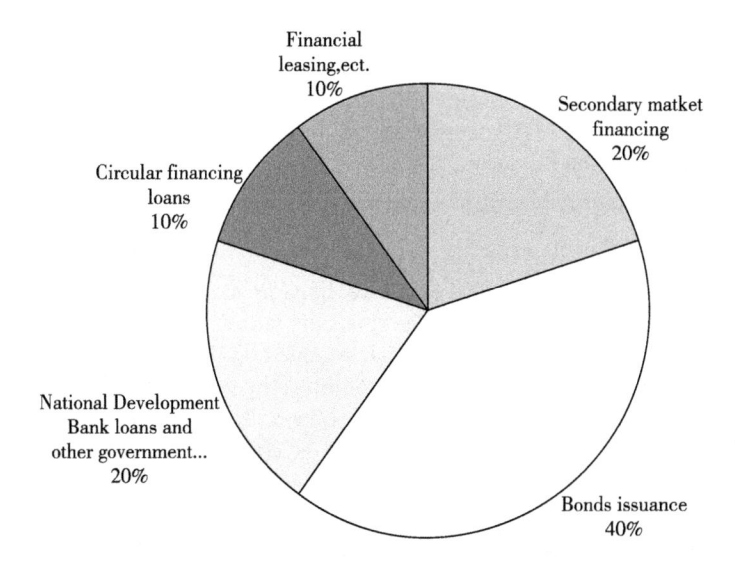

Fig. 4 Main financing structure of Aker

Financing is mainly from the traditional bank loans, diversified cruise financing mechanism is not sound, bond financing and other models are immature, and commercial banks have not established a professional cruise business financing directory.

Drawing lessons from international experience, the breakthrough path for the financial system service which serves China's cruise ship manufacturing industry includes:

Firstly, give appropriate policy tilt, introduce a series of financial support policies to broaden the cruise design and construction of corporate financing channels, and establish a diversified financing model.

Secondly, extensively introduce social capital and attract private investment through government funds and tax incentives, etc. For example, set China Shipping Investment Fund, the financial institutions establish a special financing list for the cruise industry companies and release the restrictions of the pension, insurance companies on the cruise industry.

Midstream of the Industry Chain (Cruise Ship Operation):
Bohai Ferry and CCL

Contrastive Analysis of Key Financing Modes

The financing composition of Bohai Ferry[3] is: commercial bank loans (including mortgages, guaranteed loans, etc.) accounts for 50%, secondary market financing accounts for 30%, other financing accounts for 10% and bond financing is almost zero (Figs. 5 and 6).

The world's first cruise operator CCL (Carnival Group) is the international benchmarking enterprise.[4] It has diversified financing structure: bond financing accounts for the largest proportion (50%, including private placement bonds and public offering bonds) and government policy financial support also expands to export credit (10%) and other models.

Breakthrough Path

From the experience of developed countries, bond financing is the largest source of funds for foreign cruise operators. The financing structure of China's cruise operators is mainly confined to the traditional commercial bank loans. Thus, it is difficult to meet the strong financing needs of cruise operators.

Drawing lessons from international experience, the breakthrough path for the financial system service which serves China's cruise ship operation enterprise includes:

Firstly, the domestic cruise operation company should pay attention to the importance of bond financing, and actively expand the financing channels.

Secondly, actively improve China's bond market and establish multi-level bond market to provide the cruise enterprise with better service; appropriately expand

---

[3] According to the summary of Bohai Ferry's annual report, the percentage is estimated.

[4] According to the summary of CCL's annual report, the percentage is estimated.

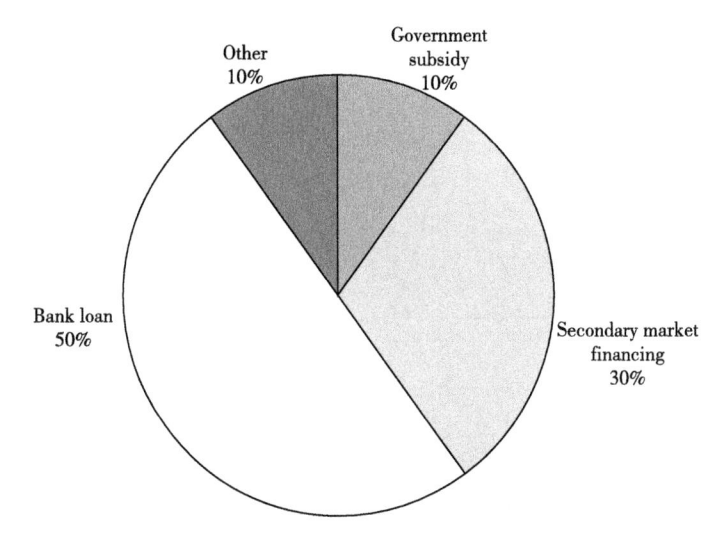

9 Research on the Financial Service System for Development ...   207

Fig. 5 Main financing structure of Bohai Ferry

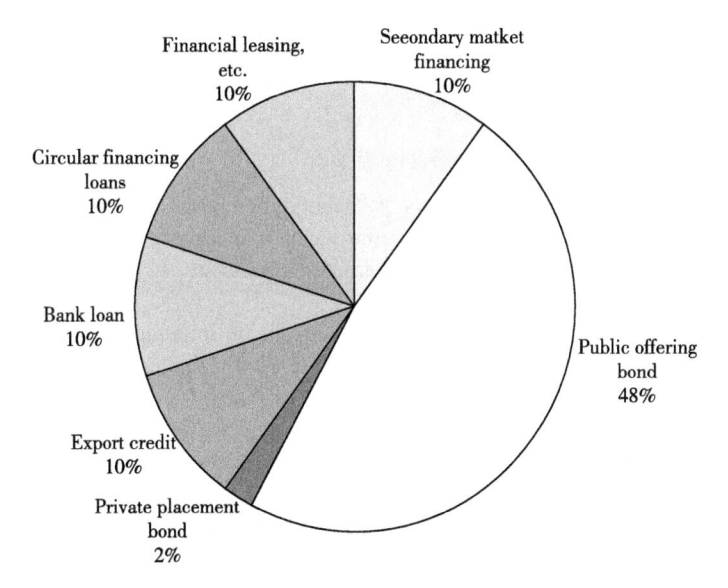

Fig. 6 Main financing structure of CCL

corporate credit bond issuance scale, reduce administrative intervention to improve market efficiency; promote the development of high-yield bonds and other innovative financing methods; promote the development asset securitization and the large shipbuilding enterprise sets up SPV as the sponsor.

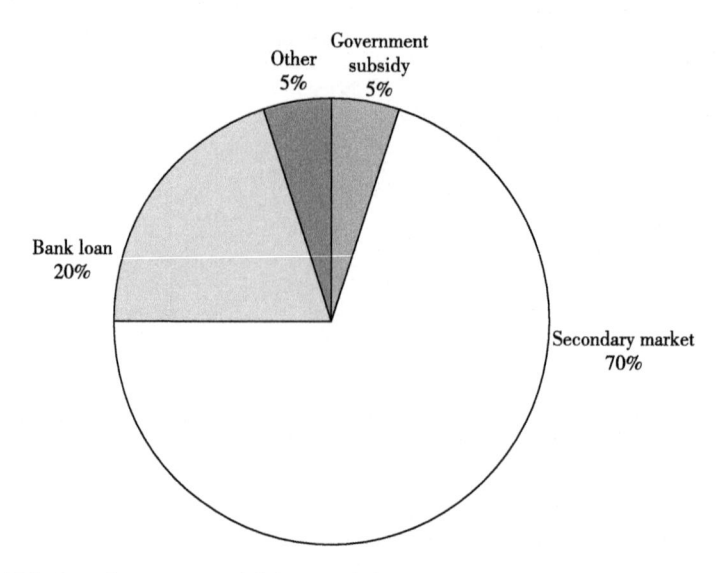

Fig. 7 Main financing structure of Caissa touristic

Downstream of the Industry Chain (Tourism Service): Caissa Touristic and TUI

Contrastive Analysis of Key Financing Modes

Caissa Touristic[5] is financed mainly by secondary market financing (70%), followed by bank loans (20%, including mortgages loans, pledged loans, guaranteed loans and credit loans) and less government subsidies (5%), In general, the financing structure is relatively simple.

TUI, the world's largest travel service group, is an international benchmarking company.[6] It has diversified financing structure. Firstly, financial leasing accounts for the largest proportion (30%). A large part of aircraft, cruise ships and hotels under the Group is carried out through financial leasing. Secondly, bank loans are also important financing channels (25%). Thirdly, in terms of bond financing, it also reflects the diversification trend, including convertible bonds and other forms (Figs. 7 and 8).

Breakthrough Path

From the international experience, diversified financing and exchange rate risk management are important financial means to support the development of cruise tourism service industry.

---

[5]According to the summary of Caissa Tourism's annual report, the percentage is estimated.

[6]According to the summary of TUI's annual report, the percentage is estimated.

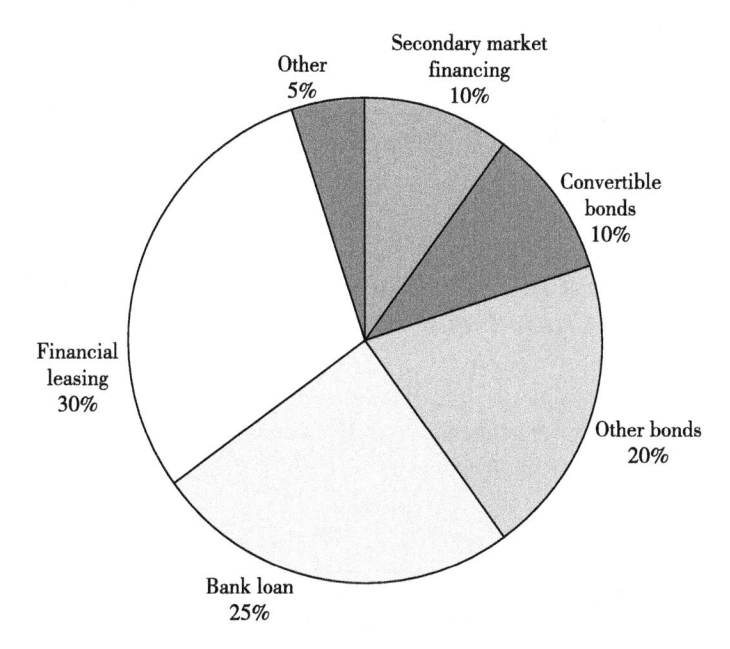

Fig. 8 Main financing structure of TUI

Drawing lessons from international experience, the path for the financial system which serves China's cruise ship service enterprise includes:

Firstly, actively expand the financing method and strengthen exchange rate risk management. As the downstream service of the cruise involve international consumer business, cruise service companies must increase risk management measures and carry out foreign exchange derivatives transactions to achieve the purpose of hedging.

Secondly, deepen the reform of foreign exchange management, encourage foreign financial leasing companies, security companies, investment companies and insurance companies to perform business in China, providing the cruise service company with convenient foreign exchange business and a diversified financing environment.

Summary Based on the Domestic and Foreign Comparison

Compared with the international benchmarking enterprises, the financing mode of enterprises of the upstream, midstream and downstream of China's cruise industry is limited to commercial bank loans, government subsidies and other traditional areas. In the emerging financial service, it is relatively weak or even missing, and there is shortage in the financial supply side, which has become an important bottleneck restricting the development of China's cruise industry.

Drawing lessons from international experience, the important breakthrough path for China's financial service system to better serve the cruise industry is to make the policy finance and market finance go hand in hand, implement diversified financing, strengthen the financial leasing, promote convenient foreign exchange and risk management.

2.4 International Experience of Financial Service System of Cruise Industry

From the international perspective, the financial service system that supports the development of the cruise industry can be divided into the European model, the US model and the Asia-Pacific model.

2.4.1 European Model

As world cruise manufacturing industry is mainly monopolized by some developed shipbuilding industry in Europe, 80% of the cruise ships are produced by the four European giants. This is inseparable from its competitive cruise financial service system.

The European model is represented by countries such as Germany, France, Italy, Finland and Spain. These countries have advantages in cruise design, development and construction. Their financial service system is characterized by the organic combination of government finance and market finance, as well as strong financial policy support of government industry in addition to traditional financing, bank loans and other financing methods.

(1) Government industry subsidies and financing offers: The European Commission has jointly issued the *LeaderSHIP 2020 Strategy* with the shipbuilding and marine industry of the EU countries, under which the shipbuilding industry is the only industry in the EU to enjoy industry-wide subsidies.[7]
(2) Revolving credit loan of financial institutions: Revolving credit loan is a flexible financing model, by which the financial institutions commit to providing loans in a certain period of time, and the borrower can use or return the loan in advance by paying a certain commitment fee. For example, Norwegian Cruise Line Holdings Ltd. was granted up to USD 610 million revolving credit loans in 2006.
(3) Shipping fund financing model: This model is represented by the German KG Fund, in which at least one partners bear unlimited liability, and the others assume limited liability as limited partners. It is generally closed-end funds. The shipping fund owns the ship and leases it to a cruise line. The cruise line

---

[7]From the *LeaderSHIP 2020 Strategy* jointly issued by the European Commission and the shipbuilding and marine industry of the EU countries.

itself has very little investment. After the new ship is put into service, its proceeds can be used to repay the principal, interest and dividends. At the same time, the ship is off-balance sheet assets for the cruise line, thereby reducing the asset-liability ratio.

Due to the combination of government finance and market finance, the European cruise industry has an unparalleled advantage in the upstream cruise design and construction.

2.4.2 U.S. Model

The United States occupies a leading position in the cruise operations and home port construction of the midstream cruise industry chain. The world's top two cruise groups are the United States' Carnival Cruise Line and Royal Caribbean Cruises. At the same time, the United States also has the world-renowned cruise ports—Port Miami, Port Canaveral and Port Everglades.

"U.S. model" is characterized by the use of mature financial markets, the expansion of the deep industry chain, and the improvement of the value chain level of the cruise industry. "U.S. model" gives full play to the role of capital markets and consumer finance: In the upstream cruise design and manufacturing link, highlighting the role of capital markets, especially venture capital; in the midstream cruise operations, highlighting the role of financial leasing; in the downstream cruise tourism consumption link, highlighting the role of consumer finance.

(1) Capital market financing: Capital market financing is more efficient and convenient than bank financing. Private equity and hedge funds play a key role in the development of the shipping industry, such as the acquisition of a large stake in Floatel Maritime and York Capital Management by Oaktree Capital Management for joint venture; the investment of Carlyle Group in Seaspan Corp. At the same time, many cruise operators also listed financing process, such as Carnival Cruise Line and Royal Caribbean Cruises.
(2) Financial leasing: Financial leasing is a new type of financial model, in which the lessor buys a cruise ship and leases it to the cruise line, who will pay rent in installments. After the expiration of the rent, the cruise line buys the cruise ship at the agreed price. In the 1950s, the United States has carried out the ship financial leasing business, and developed supporting laws and regulations, with about 5% average annual growth rate of fixed assets investment, and up to about 20% equipment financial leasing growth rate.
(3) Consumer finance: The convenience of consumer finance facilitates the consumption upgrade of US cruise tourism economy. The consumer finance is developed in the United States, and consumer credit varieties are extremely rich, which can fully meet the diversified credit demand of cruise passengers; at the same time, the financial service system in major ports supports the free currency conversion and facilitates the cruise tourism.

Therefore, the strong financial service system in the US supports the huge capital operation demand in the middle-to-downstream cruise industry, and enhances the value chain level of the cruise industry.

2.4.3 Asia-Pacific Model

Asia-Pacific region is the fastest growing region in the world cruise economy in recent years. The Asia-Pacific model is represented by Singapore and Hong Kong, China. It is characterized by the government's policy of actively encouraging the development of cruise economy, the provision of tax incentives and guidance, and playing the advantages of offshore finance. Policy finance and offshore finance form a financial force to promote the development and upgrading of the cruise industry.

(1) Singapore

Offshore financial support and government financial guarantee. On the one hand, it is the offshore financial support. Offshore account is separated from onshore account, allowing a certain degree of penetration between the two accounts. It is flexible and avoids risks for the investment and financing of the cruise business; floating exchange rate is implemented with developed foreign exchange settlement business. On the other hand, it is the government financial support. Singapore has set up a dedicated cruise financial service agency—the Singapore Cruise Development Agency and the USD 10 million fly-cruise development fund to enhance port carrying capacity and to help upgrade the cruise industry chain. Singapore is also a world-recognized low-tax country. A series of tax incentives has been made for the development of the cruise industry.

(2) Hong Kong

Offshore financial support, government financial guarantee and regional synergies. The first is offshore financial support. For medium- and long-term capital borrowings, there is no strict application procedure for Hong Kong offshore cruise lines, and there is no need to set up a separate offshore account. For operations with the onshore accounts, there is no restriction on the in-and-out funds. At the same time, Hong Kong is a free, safe and low-tax financial center. Its insurance business ranks first in the world. It has a perfect cruise insurance business, which facilitates the development of the cruise downstream services.

The second is government financial guarantee. The cruise industry is also one of the pillar industries in Hong Kong. The government gives financial support for the cruise industry. For example, the Hong Kong Tourism Board launched a new cruise fund in 2016 for the development and marketing of cruise tourism products. The third is regional synergies. The development of Hong Kong cruise industry is inseparable from the regional synergies, such as the cooperation of Genting Hong Kong Limited with China Shenzhen Merchants Shekou Industrial Zone Holdings for the construction of Taizi Bay Cruise Port. Leveraging the comprehensive advantages of China Merchants in the financial field, it aims to build a whole industry chain integrated home port construction and operations, cruise tourism and so on.

Based on the above analysis, we can find that the Asia-Pacific model has the following characteristics: highlighting offshore financial support and government financial guarantee to jointly promote the development of the cruise industry and the upgrading of the value chain.

2.5 Policy Suggestions on the Development of Financial Service System in Chinese Cruise Industry

In order to effectively realize the development and upgrading of China's cruise industry, we can absorb the advantages of "European model, U.S. model and Asia-Pacific model", and form the "China model" with the development of the cruise industry supported by the financial service system.

"China model" is characterized by leaning from the experience of various international models, rendering services in all inks of the cruise industry chain, and providing investment and financing, risk management, facilitated financial transactions and other comprehensive financial services.

2.5.1 Paths of the Financial Service System Supporting the Development of Cruise Industry

Based on the "nine-grid pattern", the financial service path of all links of the cruise industry chain is studied.

Financial Service of the Upstream Cruise Industry Chain

The output of cruise design and manufacturing links accounts for about 20% of the entire cruise industry chain, with broad market space. Involved in cruise design and construction and open up the cruise industry chain is the important direction of China's cruise business development (Fig. 9).

First of all, the upstream design and development, manufacturing and maintenance links of the cruise industry are capital-intensive with major risks and other characteristics. The role of the government in guiding funds shall be played to increase investment in cruise design, R&D and manufacturing areas by attracting venture capital and social capital. At the same time, there are major risks in cruise

Financial service / Value range	Investment and financing service	Risk management service	Convenient transaction
Upstream: Design and building of the cruise ship	Venture capital, revolving credit loan, capital market financing, etc., FTZ finance	Innovative insurance (risk compensation mechanism)	FTZ finance (offshore finance)
Midstream: Construction of home port and operation of the cruise ship	Government investment special fund, PPP model, capital market financing (stock, bond, fund), finance leasing, etc., FTZ finance	Pollution liability insurance, water traffic insurance	FTZ finance (offshore finance)
Downstream: Cruise tourism reception service	Consumer credit and other consumer finance	Cruise insurance (cruise ship cross-border theft, tourist insurance, etc.), financial derivatives hedge	Exchange settlement, bank card service, payment and storage, cross-border E-commerce platform, FTZ finance

Fig. 9 "Nine-grid pattern" for connotation of "financial services system for the development of the cruise industry" with Chinese model

research, development and manufacturing processes and the innovative insurance shall be considered to play the role of risk compensation mechanism of the first major technical equipment.

Secondly, the current development of the international cruise shipbuilding industry has shown three characteristics, i.e. "large-scale ship, diversified internal function, huge cost of construction". These three major development trends produce a huge demand for financial services. The financial institutions shall also continue to innovate, develop cruise financing catalog, design related financial products, ease the upstream financing problems; and vigorously improve the financial leasing, revolving credit loan and other methods to strengthen the capital market financing model, including the issuance of bonds and company listing.

Financial Services in the Midstream of the Cruise Industry Chain

First of all, a lot of construction funds are required in the port terminal infrastructure construction.[8] The payback period is long. Therefore, the key is to provide smooth financing channels. The government investment, "landlord port" financing, BOT financing, ABS financing and other models may be considered. Among them, the ABS financing model (i.e. assets–support–securities) is particularly suitable for the

---

[8]Related research shows that a forecasted 1.5–2.0 billion yuan is to be invested in the construction of a large and medium-sized cruise terminal with a capacity of 1–1.5 million people.

construction of home port for large cruise ship, and a variety of financing models are often adopted in cruise port construction.

Second, in the cruise operation process, huge amounts of money is required for cruise lines engaged in the sale, repair and maintenance of the cruise ship.[9] On the one hand, capital market financing is particularly important, including the encouragement of tourism equipment manufacturing enterprises through the issuance of corporate bonds, company debts, debt financing tools, project income bonds, to support the listed financing of qualified tourism equipment manufacturing enterprises. On the other hand, the security is essential during cruise operations. It is particularly important to improve the water traffic insurance mechanism.

The Financial Services in the Downstream Cruise Industry Chain

First, the important financial services of the cruise industry include export credit, financial leasing, consumer finance, cruise insurance services (cruise ship cross-border theft, passenger insurance, etc.), international settlement, etc.

Second, cruise service enterprises should focus on avoiding exchange-rate risk, and carry out hedging by the appropriate use of financial derivatives.

2.5.2 The Policy Suggestions for the Financial Service System in Supporting the Development of China's Cruise Industry

In terms of the upstream, midstream and downstream links of the whole industry chain of cruise ships, many scholars believe that a "three-step" development strategy, i.e. downstream–midstream–upstream, should be implemented.

However, this paper argues that the current global cruise industry is in the period of great reform and adjustment. Based on China's good tourism development environment, broad market, and opportunities brought about by China's development of tourism equipment manufacturing industry as a national strategy, China should be able to go hand in hand in all links of the industry chain, that is, replace the downstream–midstream–upstream strategy with the "three steps" strategy.

China should adopt a "double track path" with the government policy-oriented finance and market-oriented financial go hand in hand and complement with each other. Specifically, the following aspects can be considered: first, strengthen independent research and development through government guidance funds and venture capital in the upstream links in the industry chain; second, build cruise headquarters economy in the midstream link through the cruise industry development fund, direct financing, financing leasing, PPP mode financing, etc.; third,

---

[9]According to statistics, the cost of a large modern cruise is generally more than USD 400 million, and the annual maintenance costs are more than USD 10 million.

```
┌─────────────────┐     ┌──────────────────────┐     ┌─────────────────┐
│ Design and      │     │ Construction of home │     │                 │
│ building of     │     │ port                 │     │ Cruise tourism  │
│ the cruise ship │ ──▶ │ & operation of cruise│ ──▶ │ reception service│
│ Design          │     │ ship Construction of │     │ Cruise          │
│ and building of │     │ home                 │     │ tourism reception│
│ cruise          │     │ port and operation of│     │ service         │
│ ship            │     │ cruise ship          │     │                 │
└─────────────────┘     └──────────────────────┘     └─────────────────┘
```

Upstream
Strengthen independent research and developmentUpstream

Midstream
Build cruise headquarters economy Midstream

Downstream
Release the market vitality Downstream

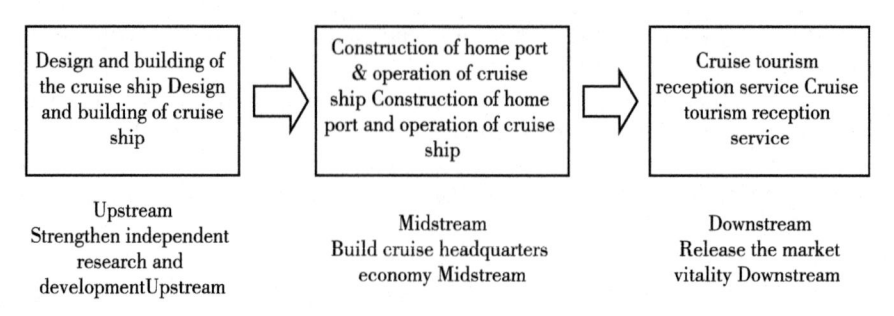

Fig. 10 "Double rack path" that supports the development of cruise industrial chain

develop offshore finance in the downstream link to play the role of the consumer finance, to build a convenient payment and settlement channel, and to fully release the market vitality (Fig. 10).

According to the above ideas, this paper puts forward the "five firsts", that is, financial innovation, policy finance, FTZ finance, green finance and "the Belt and Road", through which to further improve the financial system that supports the cruise industry chain.

Financial Innovation Goes First for the Development of Diversified Investment and Financing Services

Investment and financing services are gathered in the upstream and midstream of the cruise industry. Innovation should be made in the financial service system by learning from the advanced experience of the Europe, the United States and the Asia-Pacific region, innovative. We continuously improve China's multi-level bond market. We are committed to developing financial products such as venture capital, revolving credit loan and shipping fund. We will vigorously carry out financial leasing, build-operate-transfer (BOT), public-private partnership (PPP), asset-backed securities (ABS) and other new financing modes to expand the cruise manufacturing enterprises, home port construction and cruise line's financing channels to attract foreign investment and private capital into the cruise industry.

In the downstream cruise industry chain, we strive to develop consumer finance to improve consumer credit varieties and meet the diverse needs of tourists for the funds. At the same time, we aim to strengthen exchange-rate risk control through hedging and other financial instruments.

Policy Finance Goes First to Increase Financial Support

The government should increase financial support and develop industrial development regulations in line with the international cruise business development through financial innovation and system security. On the one hand, from the perspective of policy financial support, it is proposed to support the finance and tax policies of local cruise business with the establishment of the National Cruise Development Fund. On the other hand, it is recommended to establish a complete standard cruise special financial directory to encourage international cruise lines, international cruise organizations or industry associations in mainland China to set up regional headquarters and to create a favorable investment and financing environment for the development of the cruise industry.

At the same time, it is proposed to carry out cruise tourism property rights transactions, build trading platform for the cruise tourism projects and capitals, provide entry and exit channels for venture capital, entrepreneurship capital and social capital and escort financial institutions to serve the cruise industry.

FTZ System Innovation Goes First to Vigorously Develop Offshore Financial Services

Shanghai, Tianjin, Xiamen and other cruise cities are also FTZ pilot cities. These cruise cities should give full play to the financial advantage of FTZ pilot. Such as allowing enterprises registered in the cruise port area to open an offshore account and realize the currency free exchange in part of the tourism consumption in the cruise port when the conditions are ripe to create a good international settlement exchange and credit environment for the development of local cruise economy.

Green Finance Goes First to Guide the Cruise Green Economy

In the development of the cruise industry, the increase in emissions and wastes inevitably leads to the destruction of the local environment, so the green development of the cruise economy is very important.

On the one hand, it is suggested that the government should formulate relevant policies to guide the development of green cruise economy and to reduce the carbon intensity of cruise ships and the discharge of waste water and waste gas in combination with the *Guiding Opinions on Building Green Financial System*, *Green Ship Code* and low carbon economy ideas.

On the other hand, it is proposed that financial institutions should increase support for technological research and development of cruise enterprises in accordance with the *Green Credit Guidelines*; timely introduce green bonds, green trust and other financial products related to the cruise industry, to guide the flow of funds to environmentally friendly cruise business, maintain the ecological balance of land and sea and realize the sustainable development of cruise economy.

"The Belt and Road" Goes First to Promote China's Cruise Industry to Go Out

"The Belt and Road" initiative is a strong driving force for the business development of Chinese cruise industry. Under the leadership of the "the Belt and Road" initiative, many national and local policies are formulated to draw attention to and give strong support for the development of cruise economy. This synergy effect will promote the Chinese cruise industry to go out.

The cruise industry is characterized by long cycle of payback and large financing needs. The public financial products launched by various countries and international financial institutions have provided strong support for the Chinese cruise industry to go out. Such as the new financing platform provided by Asian Infrastructure Investment Bank, the BRICS Development Bank, etc.; financial institutions such as Silk Road Fund may provide large-scale and long-term financing support for the upstream and midstream cruise industries. In addition, the international M&A loan, international factoring and other financial support business will also provide diversified financing support for the cruise industry.

2.5.3 Policy Suggestions on Financial Policy System to Support the Development of the Cruise Industry in Shanghai

Shanghai Develops the "New Geese Model" for the Cruise Economy

At present, Shanghai is bestowed "good timing (usher in the blowout period for the development of the cruise industry), geographical convenience (Shanghai's geological advantages, as well as Shanghai-oriented national strategic advantages) and good human relations (China has the world's largest cruise market, Wusongkou cruise port has its own unique advantages)" in the development of cruise economy. Based on the unique advantages of developing the cruise economy, Shanghai is expected to take the lead in the "China model" to form a competitive "Shanghai model". "Shanghai model" has the following characteristics: on the one hand, it plays Shanghai's unique geological advantages and ship manufacturing technology advantages; on the other hand, "four centers" are built relying on Shanghai, namely the global innovation center, Shanghai free trade pilot area, Shanghai China cruise tourism development pilot area and other relevant national strategies; by playing the integration advantages of national strategies, it aims to achieve breakthrough in the development of the cruise industry.

Reviewing the path of Shanghai to develop cruise economy, we put forward the "flying geese model". The connotation of this model includes two aspects: on the one hand, follow the development of developed countries (through the introduction, imitation, cooperation and other methods to follow the development) in the mature areas of the global cruise market (such as luxury cruise R&D and manufacturing). In these areas, a "herringbone geese array" is formed with the developed countries as the leader and China as the follower. On the other hand, in some emerging or characteristic cruise industry, China and developed countries are basically at the same starting line, and a "linear geese array" is formed among China and developed countries. Shanghai may make breakthrough in these areas and take the lead in Asia and even in the whole world by playing its good timing, geographical convenience and good human relations, (such as: in the cruise business, taking advantage of Chinese huge market and the growing influence of Chinese economy and Chinese culture, Shanghai may take the lead in the cruise business with Chinese elements; in the cruise sales model, by playing China's Internet + highly developed advantages, Shanghai may rank the forefront in the Internet + cruise business, in the cruise route area, as the intersection of "the Belt and Road" and the Yangtze River economic zone, Shanghai can lead "the Belt and Road", the Yangtze River economic zone and offshore cruise travel routes), and ultimately becomes the leading geese in these areas, and other countries as followers. Thus a "linear geese array" is formed among China and developed countries.

The "herringbone geese array" with the developed countries as the leader and China as the follower in the traditional cruise field and the "linear geese array" with China and developed countries at the same starting line to one with China as the leader in the emerging and characteristic areas are the so-called "new geese model".

It is suggested that the Shanghai cruise financial service system should focus on three aspects, namely policy finance, emerging finance and FTZ finance.

Policy finance

First of all, at the cruise design and manufacturing level in the upstream of the cruise industry chain, since cruise manufacturing is a capital and technology intensive industry, cruise manufacturing enterprises cannot afford the expensive cost of building cruise ships, so the government and related agencies need to provide appropriate funds.

Second, as with the cruise terminal construction in the midstream of the industry chain, government funding is equally important. Government-supported cruise industry development fund can achieve the precise support of the cruise industry. For example, Singapore established a SGD 10 million cruise fund (FCDF Fund) in 2006, which is co-founded by the Singapore Tourism Board, the Singapore Civil Aviation Authority and the Singapore Cruise Center. The Fund is mainly established to encourage cruise lines to develop products, reward the travel agency that solicits group tours and increase the cruise market promotion efforts. For example, in order to build a cruise center in the Asia-Pacific region, the Hong Kong Special Administrative Region (HKSAR) has invested over HKD 8 billion to build the Hong Kong Kai Tak Cruise Terminal to promote the development of Hong Kong's cruise tourism industry and Hong Kong's economy. From the experience at home and abroad, since huge costs are required for the construction of the cruise terminal, so a lot of cruise port cities provide resources to support the construction of the home port. It is proposed to set up the cruise industry special fund in Shanghai to support the construction of Wusongkou home port.

Third, the government funds have an important guiding role, which will facilitate the building of the cruise economy investment and financing platform. On the one hand, we can consider the construction of government-led and market-oriented cruise industry investment and financing platform to attract foreign and domestic funds to participate in the development of the cruise industry. On the other hand, the cruise industry property rights trading center is proposed to be established to explore the cruise tourism property rights transactions and promote the close connection of cruise resources with property rights and provide entry and exit channels for venture capital and other social capital to enter into the cruise market.

Emerging Finance

As referred to in this paper, the emerging finance of the cruise industry includes both the emerging financial format and the traditional financial format that was not previously used in the cruise financial service system, such as finance leasing, capital market financing, Internet finance, cruise insurance and so on.

In the upstream industry chain, the procurement of a cruise ship with financial leasing is an important means of financial support.

In the midstream industry chain, the construction of the cruise home port requires the financial system to provide investment and financing support and risk management services. The financial support for home port can be obtained from the leasing company through financial leasing, and corporate bonds issuing, corporate debts issuing and company listing can also be selected to obtain capital.

In the downstream industry chain, financial support is required for investment and financing, risk management and financial transactions for visitors of the cruise port. Internet finance is proposed to be used for the establishment of a cross-border e-commerce platform for Chinese cruise tourism to optimize the payment function system and improve the payment efficiency and security.

Cruise FTZ Finance

From the development experience of the cruise industry in Singapore, Hong Kong and other regions, offshore finance is an important financial means for the development of the cruise industry.

FTZ is an important area for offshore finance. For example, Tianjin international cruise home port is built in FTZ, which provides a basis for carrying out better cruise financial services. Although the Shanghai Cruise Port is not established in the Shanghai Free Trade Area, Shanghai International Cruise Industry Development Comprehensive Reform Pilot should also be the carrier of system innovation practice and radiation of Shanghai Free Trade Zone. It is recommended that Shanghai Cruise Port should be fully docked with Shanghai Free Trade Area. With the help of the priorities of the free trade area and negative list management model,

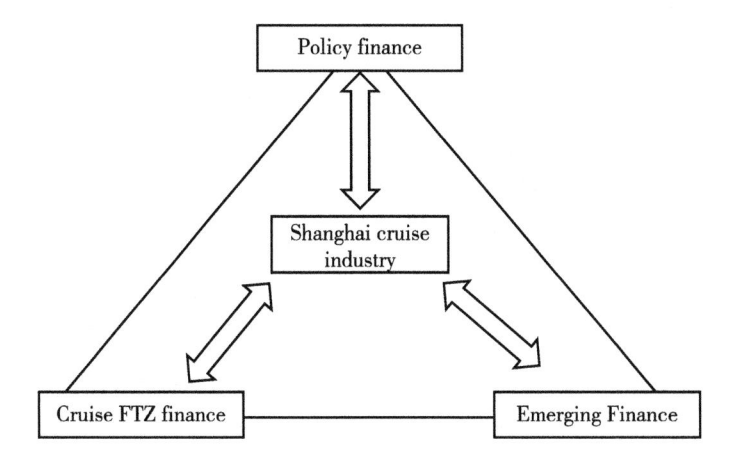

Fig. 11 Three breakthroughs of Shanghai cruise industry financial service system

system innovation should be introduced in Shanghai Free Trade Area for the construction of "Shanghai Cruise FTZ". On the one hand, the system innovation related to the cruise industry of Shanghai Free Trade Area should be copied by the Shanghai Cruise Port; on the other hand, the system innovation corresponds to the Shanghai FTZ finance should be copied by the cruise port, to create a good financial environment for the cruise port.

Breakthrough in cruise FTZ finance should be made from two aspects: first, strive to provide the cruise port enterprises the same financial services with FTZ enterprises. For example, equal treatment should be sought between enterprises registered in the cruise FTZ with those registered in the FTZ pilot in the setting of free trade account, investment and financing exchange, foreign financing and other aspects. Second, the same financial services should be provided for the cruise port visitors and FTZ residents. Such as providing cross-border consumer credit services and comprehensive services for domestic and foreign tourists, including foreign exchange and settlement, bank card, credit card transactions and payment storage.

Policy finance, emerging finance, cruise FTZ finance constitute the three important breakthroughs of Shanghai cruise industry financial service system to jointly promote the development of the cruise industry and the value chain upgrade. See details in Fig. 11.

In summary, China's cruise financial service system can take "step by step" development strategy: first, in the upstream links in the industry chain, strengthen the independent research and development and manufacturing capacity of the cruise manufacturing industry through the government cruise industry fund, venture capital, Sino-foreign joint venture funds; second, in the midstream links in the industry chain, strengthen the construction of the home port through direct financing and PPP mode financing to build the cruise headquarters economy; third, in the downstream links in the industry chain, give full play to the role of consumer finance, Internet finance and FTZ offshore finance, and fully release the vitality of the cruise market.

In the process of building a more competitive and dynamic cruise industry financial service system in Shanghai, policy finance, emerging finance, cruise FTZ finance is the three important breakthroughs.

Chapter 10
Legal Coordination Difficulties in the Course of Chinalization of Cruise Industry and Coping Measures

Fangyuan Lv, Zhaobin Pei and Jie Zheng

1 Necessity for Law Supply Side Reform in Chinalization of Cruise Industry

1.1 Legal Regulation Requirements for Chinalization of Cruise Industry

Currently, scholars do not have an authoritative and unified definition of cruise industry. In general, it refers to cruise ship-based industries of tourism, traveling, leisure, entertainment, etc. including industrial economies related to ship design, ship production, company operation, maintenance of related equipment, port services and sightseeing. From the perspective of supply side, the cruise industry chain contains upstream manufacturing and manufacturing related industries, midstream cruise lines and other cruise industry, and downstream reception and management of cruise terminals.[1] The Costa Allegra of Costa Cruise Lines operated cruise routes by taking Shanghai as its home port for the first time in 2006. Ever since then,

---

[1]Tengfei Zhao, China's Cruise Industry Development Countermeasures Based on the Perspective of Industry Chain, *Journal of Suzhou University*, 2016/Issue 6, Page 15.

F. Lv (✉)
Marine Law Institute, Dalian Ocean University, Dalian, China
e-mail: lvfangyuan2008@163.com

Z. Pei
Law School and Coast Guard School, Dalian Ocean University,
Dalian Maritime University in International Law, Dalian, China

J. Zheng
Law School and Coast Guard School, Dalian Ocean University, Dalian, China

© Social Sciences Academic Press and Springer Nature Singapore Pte Ltd. 2018
H. Wang (ed.), *Report on China's Cruise Industry*,
https://doi.org/10.1007/978-981-10-8165-1_10

a cruise ship, as the "mobile five-star hotel", has become more and more popular among Chinese people.[2] From January to May 2017, according to statistics, China's cruise ports have received 472 inbound and outbound cruise ships (year-on-year growth of 31%) and 1.687 million inbound and outbound tourists (year-on-year growth of 23%).[3] The domestic home ports have received 413 cruise ships (year-on-year growth of 29%) and 1.574 million tourists (year-on-year growth of 23%); visiting ports have received 59 cruise ships, with a year-on-year growth of 40%.[4] The clustering of cruise industry belongs to different economic sectors making law departments to adjust boundary obstacles. As it involves the supervision value of public laws and has the meaning of private laws, the freedom value of autonomous contracts shall be classified into it. There are both the scope for jurisdiction conflict between flag state and coastal state in international law, and regulation example of domestic economic law.[5] Moreover, there is application scope of *Civil and Commercial Law*, *Maritime Law* and *Tourism Law*. Macro and micro regulation makes legal regulation face the application problem. Due to the hysteretic nature of laws, the source of law of Anglo-American law system and civil law system is different, and the development stage of Chinese and western cruise industries varies too.[6] The rapid Chinalization of cruise industry may result in some specific problems such as "tourists' occupation of cruise ship", "arrest of cruise ship", "ownership of some shipping space", "ownership of all shipping space" and lagged development of cruise services, which will put forward new research trend on China's existing law regulation mechanism. In the pattern of Chinese deductive thinking, the regulation of the cruise industry presents a fragmented pattern, such as transportation regulation rules under the *Maritime Law* and tourism regulation rules under *Tourism Law*. China's existing cruise industry needs legal policy supply side reform to jump out of the set pattern of existing legal sector.

The main reason for fragmentation in management of cruise industry is the double attributes of cruise ship, i.e. the transport and tourism attributes. Due to boundary obstacles in clustering industrial law regulation, various administrative departments have developed government regulations and industrial development policies from the perspective of industrial development. Its flexibility, multi-layer and mutual conflict nature are becoming more and more apparent. Due to the double attributes,

---

[2]Mengyao Zhang and Yun Liu, Research Summary on Development of Cruise Tourism, *Journal of Baoshan University*, 2014/Issue 1: 76.

[3]Nine Cities under Construction or Key Cruise Terminals, *Pearl River Water Transport*, 2013/Issue 14.

[4]Jianrong Chen, Gradually Realize the Dream of Manufacturing Luxury Cruise Ship in China, *Guangdong Shipbuilding*, 2016/Issue 4: 12.

[5]Ying He and Xiaoming Wu, International Economic Law (Volume One), Northeastern University Press, 2006: 25.

[6]Fangyuan Lv: Research on Law of Cruise Ship under Transport Perspective, Doctoral Dissertation, Dalian Maritime University, 2015.

there are regulations of two departments to regulate the cruise ship itself, namely, the dual regulations of CNTA and MOT.[7] MOT defines the industry as per the attribute of the cruise ship without considering the tourism attribute of cruise ship. It lacks of understanding on the statement that takes cruise ship as a tourist destination. CNTA mainly defines cruise ship as per its tourism attribute and cruise ship is mainly regulated by the travel agencies who manage cruise tourism. The unfairness of standard clauses on steamer ticket and contract between cruise lines and tourists can only be supervised by SAIC.[8] Cruise industry involves entities such as cruise lines, tourists, port managers, travel agencies and cruise logistics operators, and has relationship of administrative subordination of MOT, CNTA, GACC, CIQ, etc.[9]

1.2 Practical Analysis of Legal Regulation in Cruise Industry

China's existing legal provisions on the cruise ship are basically administrative regulations issued by the State Council and various authorities or rules and regulations designated by local governments at all levels.

1.2.1 Guidance and Planning of Cruise Industry of the State Council

S/N	Document name	Main contents
1	August 2008, *Guidance for Promoting Development of China's Cruise Economy*	The basic principles, overall objectives and main tasks of the Guidance mentioned: construct legal and policy service system, and protect the development of the cruise industry[a]
2	October 2009, *Notice on Improving Management of International Navigation Ship Port Supply Market*	The Notice plans[b] the cruise industry mainly from the perspective of the cruise services, analyzes the opening of supply market mainly from the economics perspective of port supply market, actively explores the new model of sound development of cruise supply industry and constructs supporting policies and measures

(continued)

---

[7]Fangyuan Lv, Research on Law of Cruise Ship under Transport Perspective, Doctoral Dissertation, Dalian Maritime University, 2015.

[8]Fangyuan Lv and Ping Guo, Cruise Economy Analysis under the Perspective of Law, *Social Sciences Review*, 2014/Issue 1: 69.

[9]Yongping Tong, Railway-river Combined Transportation Enters a New Age in China's New Round of Development, *New Silk Road Horizon*, 2013/Issue 1: 44.

(continued)

S/N	Document name	Main contents
3	December 2009, *Opinions on Accelerating Development of Tourism*	Plan the cruise industry mainly from the perspective of tourism industry[c] ① Include cruise ship manufacture into the national encouraged industrial catalogue ② Stipulate to take the development of cruise ship and yacht as a new tourist consumption hot spot
4	February 2013, *National Tourism and Leisur e Outline (2013–2020)*	The Outline clearly puts forward to actively support the development of cruise and yacht industry in infrastructure, and tourism and leisure products
5	In March 2015, the Belt and Road initiative officially announced the cooperative focus, cruise tourism	① Enhance cruise tourism cooperation and improve visa service facilitation of countries along the Belt and Road to achieve cruise tourism sharing economy ② Construct cruise tourism as an important part and node of the 21st Century Maritime Silk Road[d]

[a] Jiaxuan Gao, Shanghai Cruise Policies and Laws Research, *Chutian the rule of law*, 2015/Issue 7: 35
[b] Xiaonian Li and Chenguang Yan, Some Policy and Legal Issues in Development of China's Cruise Industry, *Chinese Journal of Maritime Law*, 2013/Issue 3: 48
[c] Feng Su, Analysis on Risks in Development of China's Cruise Tourism and Coping Strategies, *Journal of Wuhan Communication Management Institute*, 2014/Issue 1: 51
[d] Manli Peng, Creating a Legal Environment for Marine Economy—with Cruise Tourism as an Example, *New Economy*, 2014/Issue 31: 25

1.2.2 Cruise Transport Provisions of the Ministry of transport

At present, Ministry of transport has always paid attention to the extension of cruise transport function, and stipulates that the essence of the cruise industry is ship.[10]

S/N	Document name	Main contents
1	April 2011, *Foreign Investment Industry Guidance Catalogue (Revised in 2011)*	Clearly encourage: the design of luxury cruise ship and deepwater (over 3000 m) marine engineering equipment in transportation equipment manufacturing (limited to joint stock and cooperation), and design and manufacture of yacht (limited to joint stock and cooperation)[a]. Include the cruise manufacturing into foreign investment encouraged catalogue
2	September 2011, Notice on Strengthening Approval and Management of Wholly Foreign-owned Shipping Companies	Loosen the rules for wholly foreign-owned cruise lines to loosen the market access system[b]
3	March 2014, *Guidance for Promoting Sustainable & Sound Development of China's Cruise Transport Industry*	The development and introduction[c] of the Guidance will further clarify the future development direction of domestic cruise industry. The Guidance, however, lacks of specific operational provisions ① Provide guidance from the cruise tourism market cultivation, cruise route development and design and port planning & layout, to promote the development of cruise transport industry ② Build a sound development cruise market system from the perspective of industry regulation, and based on service and market adjustment Build complete cruise transport service standard system by taking transport services level and quality of service as entry points ③ Build cruise ship economic chain and enhance the contribution of the cruise economy by taking cruise transport as entry point

[a]Information Dynamics, *Finance & Accounting for Communications*, 2014/Issue 10
[b]Zhe Ni, Cruise Research on Legal Issues in Ship Operators' Liability Insurance, Master's Dissertation, Dalian Maritime University, 2015
[c]Fangyuan LV: Research on Law of Cruise Ship under Transport Perspective, Doctoral Dissertation, Dalian Maritime University, 2015

---

[10]Fangyuan Lv: Research on Law of Cruise Ship under Transport Perspective, Doctoral Dissertation, Dalian Maritime University, 2015.

1.2.3 Cruise Tourism Regulations of CNTA

S/N	Document name	Main contents
1	February 2009, *Waterfront Tourism Planning*	① Include cruise tourism into waterfront tourism planning system ② Put forward to take the development of cruise tourism products such as coastal products and inbound travel products as the entry point of China's cruise tourism market[a]
2	February 2011, International Cruise Port Tourism Service Code	Put forward industrial adjustment standard norms mainly from the requirements of international cruise port tourism services[b]

[a]Interpretation on Guidance for Promoting Sustainable Development of China's Cruise Transport Industry, *China Water Transport* (First Half), 2014/Issue 4: 13
[b]Fangfang Zhang and Baishou, Fang Prospect Analysis on the Development of Cruise Tourism in Qingdao, *China Water Transport* (Theoretical Version), 2006/Issue 6: 24

1.2.4 Establishment of Relevant Systems by Shanghai Municipal Tourism Administration

Documents	Main contents
August 2015, *Shanghai Cruise Tourism Contract Model*	(A) The model text classifies cruise lines as performing aids under the *Travel Law*. In response to frequent cancellations and changes of voyages due to weather conditions such as typhoons and heavy fog on Asian routes, the model text places particular emphasis on performance due to force majeure[a] or travel agencies Auxiliary person has made reasonable due diligence Unavoidable event causes travel to change or cancel part of the port of call, the travel agency should refund part of the travel expenses to travelers and unpaid harbor port and shore sightseeing costs. Expect to better regulate the relationship between tourists and cruise ships and travel agencies
	(B) The model text of Shanghai cruise tourism contract stipulates the rights and obligations of both tourists and travel agencies according to the *Travel Law*[b] ③ The model text of Shanghai cruise tourism contract stipulates the rights and obligations of both tourists and travel agencies according to the Travel Law.
	(C) The contract has explicitly stipulated settlement ways of cruise tourism disputes and tourists' occupation of cruise ship. If there are disputes during traveling by cruise ship, tourists shall resolve disputes as per the settlement ways specified in the contract, and shall not harm the legitimate rights and interests of others, or otherwise shall be liable for the loss

(continued)

(continued)

Documents	Main contents
June 2016, *Shanghai Cruise Ship Operation Norms*	The norm mainly focuses on cruise tourism industry regulation in terms of cruise tourism definition, business qualification, insurance, ticket sales, cruise contract, route change, facilities and service standards, special disclosure obligation and dispute resolution.
	(A) According to Article 18 of the "Operational Norms", the first to the third paragraphs of "Dispute Resolution" mainly stipulate the respective subjects of dispute resolution according to the cruise tourism products and the causes of the disputes[c]
	(B) The *Norms* define "cruise tourism" from its transport and tourism attributes[d]
	(C) The highlights of the *Norms*—in order to fully safeguard the legitimate rights and interests of consumers in China, the cruise carrier is required to configure Chinese descriptions of facilities, safety signs and instructions related to the safety of the tourists in the cruise ship
	(D) The right of the captain to change the voyage stipulated in the *Norms* is based on the international practice. Due to high requirements of maritime operation, it is internationally recognized that the right to change the voyage authorized only to the captain

[a]Linfang Fu and Hangping Que, Discussion on Relevant Issues of Package Tour Contract, *Tourism Tribune*, 2015/Issue 9
[b]Luyan Zhao, Research on Right of Terminating Tourism Contract without Any Reason, Master's Dissertation, Zhejiang University, 2012
[c]According to Article 18 of the "Operational Norms stipulates that: "In the occasion that travel agencies and international shipping agencies are commissioned to sell ship tickets, if there are cruise tourism disputes between cruise lines and tourists caused by cruise lines failing to provide relevant services as per provisions of the contract, the latter shall be liable for dispute settlement; if there are disputes caused by travel agencies and international shipping agencies failing to fulfill the duty of disclosure in ticket selling, the travel agencies and the international shipping agencies shall be liable for dispute settlement." If the travel agencies change ship tickets and shore tourism services into package tourism products violating the disputes specified in tourism contract, the travel agent shall take the lead to be liable for dispute settlement. If there are personal damage and property loss caused by cruise lines, or disputes caused by cruise ship cancellation and change, the cruise lines shall take the lead to be liable for dispute settlement and the travel agencies assist. The cruise lines, travel agencies and international shipping agencies shall actively consult with tourists to settle disputes, and truthfully inform them the ways to complain and other solutions"
[d]Cruise tourism refers to the outbound tourism that takes the cruise ship as tourist destination and mode of transport, and provides services such as maritime excursion, accommodation, transportation, restaurant, dining, entertainment or ashore sightseeing for tourists

According to the analysis on the above mentioned laws and regulations and regulatory documents relevant, the provisions of the Ministry of transport on cruise ship are for cruise transport and mainly relevant to technical aspects with few cruise tourism provisions; on the contrary, the provisions of CNTA for cruise tourism are mainly administrative supervision.

Shanghai Municipal Tourism Administration Contract Model regulates the industry from the perspective of personal profit, while the *Norms* regulates it from the perspective of market regulation under economics.

2 Legal Set Pattern in Chinalization of Cruise Industry

2.1 Legal Dilemma of Cruise Transport Industry

Legal disputes in recessivation of ship tickets. The cruise industry has transport and tourism attributes, and cruise tourism and transport industries are sold by the travel agencies in way of cruise package tourism products. That is, travel agencies package services such as ship tickets (including cruise transport, and corresponding room and catering services), shore sightseeing, visa and tour guiding, and consumers and travel agencies may sign "outbound tourism contract". Therefore, in actual operation, the status and role of tickets are less important, the responsibility of each party is unclear, and there is lacking of carrier to exercise the rights and fulfill obligations among cruise lines travel agencies and cruise tourists. In tourists' occupation of cruise ship, cruise lines believe that skipping over the port is caused by force majeure such as weather, and they do not need to bear the responsibility for breach of contract. While the tourists can not prove the cruise lines have breached the contract resulting in frequent tourists' occupation of cruise ship. In Shanghai port, hundreds of people were told that they had completed the ticket sales contract and but there was no relevant ship tickets. They also found that they did not have any information about the cruise ship nor belonged to the cruise ship before the date of embarking. It led to large-scale tourists' occupation of the cruise ship and great impact on the port security order.

Full booking rate paradoxical predicament in cruise transport contract agreement. As one of high-end tourism products, the price of cruise tourism is pretty low in China, mainly because that the cruise lines and travel agencies have stipulated the requirements for booking rate in "contracting all shipping space" and "some shipping space" (if the contractor fails to meet the booking requirements, it has to pay a fine as per the tickets not being sold). To seize market share, or based on the sales pressure of contracting all or some shipping space, the travel agencies frequently reduce the price of cruise tourism products, and sell the tickets at low price. In recent years, the cruise market, therefore, falls into the abnormal phenomenon of later ordering and lower price. The actual transaction price of a considerable amount of voyage is decreased by 50% compared with the market price, or even lower than the contacting cost of travel agencies. There are endless low rice dumping phenomena. All of those have violated the market competition order. Although almost all tickets of cruise ship will be booked before voyage, and cruise lines can both ensure the ticket revenue and obtain the upstream income of the cruise ship, most of the travel agencies are in the red. It is difficult to regulate the monopoly of cruise lines from the perspective of *Anti-monopoly Law of the People's Republic of China*. The *Law* mainly regulates two parties by contract, and the standard clauses only mention the unfair attribute in arbitration of two parties. However, two parties have been informed of the corresponding obligations when signing the contract. To avoid the risk of tickets that can not be sold out in stern room, the travel agencies sale the tickets by low price under freedom of contract. The high-end cruise tourism becomes cheap because of the travel agency's low price competition. As vicious competition below the ship ticket has become common, the threshold effect of ship ticket price has disappeared and high-end features basically destroyed, with difficulty to popularize basic cruise etiquettes and rules. And behaviors of violations of cruise etiquette such as going to restaurant in slippers and bathrobes, taking instant noodles on the cruise and fighting for food while eating buffet have seriously affected the experience of cruise tourism.

As sellers (operators) are the applicable subjects of the *Anti-unfair Competition Law*, the cruise lines, under chartering, have transferred the pricing power, and are difficult to control the actual selling price therefore. Moreover, the ship ticket seller is cruise lines rather than the travel agency with pricing power, resulting in applying anti-unfair competition law with difficulty. And the government management department has no way to set government guidance price so as to avoid monopoly.

2.2 *Legal Dilemma for Alienation of Chinalization in Cruise Tourism Agency Model*

For agency mode of travel agency, it is difficult to be approved at primary level and supervised at secondary level. The agency mode is a win-win mode of reference between cruise lines and travel agencies when it first emerged, and will be continued for a long time. The existing cruise tourism is outbound one, as the cruise lines are unqualified for outbound tourism and must rely on the travel agencies, who implement the "all in one package" of cruise products (ship ticket and shore tourism). According to provisions in the *Tourism Law*, the travel agencies shall obtain corresponding qualifications to sell package tourism products of outbound tourism. And it mainly belongs to the approval system, and the requirements on corresponding qualification are relatively high, therefore, the number of such agencies limited, with more than 130 agencies only in Shanghai. Being subjected on outbound travel agencies and in the face of rapid development of cruise tourism, the number of outbound travel agencies has imposed restrictions on the expansion of distribution channel of cruise tourism. As travel agencies implement the approval system and the market demand exceeds supply, the subordinate travel agency subcontractors emerged, which is not qualified for outbound tourism. In the name of distributors, moreover, they are engaged in selling outbound tourism products, which leads to the good and bad mixed together, and with the emergence of Internet direct booking, the market management of subordinate distributors is in a state of chaos, and subordinate distributors of travel agencies are basically in shortage of legal system. The risk of sales and operation of cruise tourism products has increased at varying degrees. For ship ticket sales, according to the market access system stipulated in the *Norms*, sales agents of cruise tourism ticket are required to obtain corresponding qualifications, which cannot be obtained by general travel agencies, the travel agencies, however, have packaged qualifications into products for together sales, leaving the provision impracticable. Moreover, the approval system for secondary agent qualification of ship tickets sales has been changed to the association for the record, and the record are canceled finally. In the context of "Internet plus", the supervision on agents of ship ticket sales has become more difficult. Under the agent mode, contracts are mainly signed between the travel agencies and tourists, and the list of tourists is not submitted to cruise lines until boarding day comes (before 45 days), resulting in the back-to-back relationship between cruise lines and tourists. And the tourists do not know whether they have bought ship ticket until boarding day. The reason for the dispute lies in that market supervision can not be started, which brings supervision pressure to local governments. Agent mode has caused difficulty to approve primary travel agencies, with difficulty to supervise subordinate distributors.

The shore tourism alienation of travel agency cruise tourism. As mentioned above, for either chartered cruise or exclusive sales at cabins, cruise products are mainly sold in the form of "cruise package tourism products" of the travel agency, and group shore tourism organized by the travel agency is constituent part of package products. In accordance with the practice of China's tourism industry, travel agencies organize tourists for shore tourism, which basically turned into a shopping tour. And tourists are forced to disembark, therefore, commission returning due to shopping while shore tourism has become the profit point, which greatly reduces the quality of cruise shore tourism. Although there are stipulations in Paragraph 2 to Article 35 of the *Tourism Law*: when organizing and receive tourists, travel agencies shall not designate specific shopping places, or provide tourism services that require additional payment, it does not include circumstances where both parties have agreed or the tourists have requested such agreements and no influence is caused on the itinerary of other tourists. Because of the tariff, however, the tourists themselves may also change their tourism into a shopping tour, even if they are not forced to shop for consumption, and the existing *Tourism Law* is difficult to regulate such behaviors. Ultimately, it was the weather that leads to skip over the port and then forcibly occupy the cruise ship, which is due to happen, because the tourists take the cruise for outbound tourism, with the true meaning of industrial tourism completely alienated. Based on global tourism, cruise lines regard the cruise ship itself as a tourist destination, while tourists only see it as a means of transport, and the cognitive gap between the two has brought design and planning risks for the Chinalization of cruise industry. At the same time, the forced disembarking of cruise tourism brings legal risks to the consumer protection of tourists, which will infringe the legitimate rights and interests of part of tourists who are not willing to disembark and which does not meet international practice.

The meaning of departure tax refund now faces legal alienation. Since the four routes of home port established in 2006 by cruise lines in China, the outbound tours by cruise ship (representing by Shanghai) has developed rapidly, the economic effect brought with it, however, is limited. One of the important factors is that cruise tourists complete a series of shopping activities and consumer behaviors on the cruise or outbound, resulting in mass consumption outflow. Data show that in 2016, the outbound passenger volume of Shanghai home port cruise enjoys a year-on-year growth of 73.46%, while the number of people who buy duty-free goods only 18.39%, which is caused largely by the large passenger volume at duty-free shops in the short period of departure time, resulting in that tourists have insufficient time for consumption and they cannot buy duty-free goods with another entry. And this is an inherent defect of departure tax refund system. If cruise tourism of overseas tourists continues to stop in other domestic ports, here comes repeated and complex operation of departure and entry then. Moreover, it is difficult to play the trend of driving the integration at cruise home port area, as security check on cruise ship itself will take a lot of time, and entry and exit procedures once again are cumbersome, which damages the actual stimulation of cruise tourism. The purpose of cruise ships' calling at a Chinese port is to bring economic benefits, but the existing departure tax refund is unclearly defined in laws, seriously affecting the shopping enthusiasm of tourists, and making the economic stimulation of cruise industry be greatly reduced.

2.3 Stiffened Legal Norms of Cruise Logistics Industry

The legal definition on cruise services is unclearly, and the multiplier effect to ports is relatively low. The stimulation of cruise industry to port economy is mainly from consumption on materials and food on a cruise ship as well as ship maintenance and building of ship accessories and even cruise ships. These are collectively referred to as cruise services. As a piece of cake in the cruise industry and in its Chinalization, the cruise industry does not give full play the corresponding economic effect of cruise ships. The profitability of home ports of cruise ships is relatively poor, and its profit point is only from the collection of port taxes and fees, accompanied by stability pressure, environmental management pressure and other port burden and problems caused by tourists' occupation of cruise ships. This does not meet the original intention of the cruise industry. Compared with general ship services, the characteristics of cruise services are different, and the general ship services only meet normal operation of the crews and ships. The cruise services are similar to the overseas purchase and shall belong to international trade, but have not been included in the international trade[11] so far. And if it is still regarded as common ship

---

[11]Fangyuan Lv, Research on Law of Cruise Ship under Transport Perspective, Doctoral Dissertation, Dalian Maritime University, 2015.

services, the product price of cruise services will be higher than that of overseas ports with same location. As there are no export rebates, the cruise services, as blocking trading goods, do not enjoy competitive advantage in price, let alone that we have food safety problem. Based on all mentioned above, the cruise services do not give full play corresponding advantages.

Law enforcement difficulty in cruise services management. In accordance with the provisions of purchase rules of international cruise lines, the cruise line headquarters shall supply ships according to the needs of fleets and via home ports. And based on international practice, the headquarters of international cruise lines, according to the needs of fleets, generally transport to port on their own and load then and transport to the port call of cruise ship' home port by plane or ship, and directly deliver to the cruise ship after passing customs declaration and inspection in form of customs transit for ship supply, with relatively fast procedures. The current customs transit of existing import food is cumbersome, however, with complex procedures, and with "customs clearance form" required as well after passing inspection and quarantine. Moreover, the properties of food determine that requirements on the efficiency of its import and export procedures are relatively high, which are inconsistent with the existing management pattern and way of law enforcement. After passing inspection and quarantine and going through the customs, load food containers purchased from South Korea or the United States on the ship. According to relevant provisions, however, food and containers are all transit goods, and containers shall be loaded on a cruise ship. This is not in line with business practice. There are no supporting legal provisions on empty container handling, and the laws are only enforced with stiffness and with the border management as the basis, seriously hindering the home port advantages of cruise logistics industry. As the tariff-guarantee and customs-transit supply on food and other materials of cruise ships is one of the most basic functions of a cruise home port, if the international customs-transit operation on materials of the cruise ship can not be normalized, and the other amount of food provided by the cruise ship will be reduced correspondingly. Some benefits supplied by cruise services will be taken away by adjacent ports in Korea and Japan. As ports in Korea and Japan are far from the home ports of our cruise ship, with uncertain factors increased, once skipping over the port is occurred due to circumstances beyond our control, great inconvenience to cruise services will be brought then.

Flag state dilemma of cruise ship. The profit point of cruise services lies in cruise maintenance as well. Currently, we are not only facing technical obstacle and talent barriers, but also legal impediment in terms of cruise maintenance. Furthermore, garbage disposal after cruise maintenance faces legal dilemma of "foreign garbage" disposal in the existing law. In accordance with the current legal provisions of MEP, since September 2017, products of cruise maintenance have been belonging to "foreign garbage" category and are prohibited to import, and wastes caused by cruise maintenance are required to carry back to home cruise for disposal therefore. And high extra cost and inconvenience will be generated in accordance with the existing laws. Hence, a lot of cruise ships choose to repair at neighboring ports in South Korea and Japan rather than their home ports in China.

2.4 Deficiency of Legal Norm in Cruise Services

There is legal deficiency in creating local fleets of cruise ships. Fleets of cruise ships as the main development of the cruise market, the economic additional value contributed by cruise lines accounts for more than half of all cruise economy, while that from upstream manufacture of cruise ship for about one-fifth.[12] At present, the development of China's cruise lines still remain at stages of purchasing foreign second-hand cruise ships or cooperative operation with foreign investment in the global cruise industry, with the flag of convenience hung in the course of operation. And the fleet is small in size, with very limited competitiveness.[13] Shipbuilding, as the last piece puzzle of the shipping industry, has been dominated by Italy, Germany, France and other shipyards in Europe. China's shipbuilding and maintenance technology is still in a blank stage, with Achilles heel of legal risk prevention.

For the industry chain, the construction of local cruise industry is slow, route design of cruise ships is single, and the laws on positive opening of new routes are rigid. The planning of cruise routes is still a part of traditional passenger transportation. With the opening of new routes, the tourism of cruise ships to high seas faces the shackles of rules and regulations of Ministry of Public Security, and the obstacles of tourism of cruise ships to high seas lies in the ban and regulation on gambling. The opening of routes-coastal and offshore routes face the legal impediment of market access between international passenger transportation and coastal transportation in the maritime.

The legal status of port operators is unclearly defined. Driven by cruise industry, coastal areas start one after another the construction projects of cruise home port, with the lack of unified cruise industry development planning,[14] resulting in serious redundant construction and homogeneous competition at the port. Moreover, there is no coupling effect of cruise economy between home port and ports of call formed. In terms of service standards of cruise ports, there are no uniform standards in establishing the standard system for reception services. And the legal status of port operators, as unclearly defined, will face a vacuum state in the legal regulation.

---

[12]Liming Tong, Ran Leng and Peng Gao, Discussion on Several Issues in Qingdao Home Port Construction, *China Water Transport* (Second Half), 2016/Issue 5: 60.

[13]Chaoyang Tan, See Ships under Flags of Convenience from the View of Fleeting Operation, *Navigation of Tianjin*, 2002/Issue 1: 35.

[14]Feng Su, Analysis on Risks in Development of China's Cruise Tourism and Coping Strategies, *Journal of Wuhan Communication Management Institute*, 2014/Issue 1: 51.

3 Chinese-Style Solutions to Cruise Tourism Industrialization

3.1 Set the Existing System as the Blueprint to Do Well in Top-Level System Design

Firstly, set the *Cruise Ship Operation Norms* as the blueprint and introduce government's regulations and exemplary legislation on cruise home ports. The *Norms* is the only official normative document of the management adjustment on existing cruise tourism, and solves the definition of cruise ships' transportation and tourism attributes in cruise industry. With the *Tourism Law* as a framework, under the agent mode of travel agencies of cruise tourism and in the face of disputes, it details how to settle disputes on program (namely, the issues of lead agency) in a better way. Therefore, the method that can be guided is acquired in private interest relief and from the management point of view. However, as the legal hierarchy of the *Norms* is relatively low, and it is required to be further lifted as rules and regulations of Shanghai municipal government, so as to provide exemplary legislation for other home port cities. Under the agent mode of cruise travel agency that can be defined by such regulations, the legal status of ship ticket agent will be clarified, and the legal system of ship ticket agent will be established to define its qualification requirements. Institutional system related to air tickets may be referred to, including qualification authentication, certificate awarding, fee special account establishment, and improvement of guarantee, insurance and other mechanisms. The real-name system of ship tickets for boarding will be implemented. After the execution of contract of cruise tourism, the ship ticket agent shall provide real-name tickets issued by cruise lines on their system platform. Tourists, before boarding at the terminal, shall handle procedures for boarding with their valid travel documents and tickets. And the normative provisions on the formation, effectiveness and dissolution of contract between both parties shall be stipulated. Define requirements on cooperative mechanism of operation among cruise ports, and actively give full play the regional competitive advantage to form competition and cooperation relationship among port cities, achieving win-win situation therefore.

Secondly, do well in top-level design of the *China's Cruise Industry Development Planning Law*.[15] In this regard, we can learn from the provisions in the *Korean Cruise Industry Promotion Law* to break the boundary obstacles on legal regulation of the existing laws. Set definitions on legal relations between subjects of the cruise industry, legal nature of cruise services and legal status of port operators, with formatted text for cruise shipbuilding provided and introduce a package of policies for cruise industry development planning for good Chinese-style top-level design of cruise industry adjustment.

---

[15]Yongguang Ma, Way for China to Develop its Own Cruise Fleets, *World Shipping*, 2013/Issue 8:19.

3.2 Set the FTZ and Experimental Area for Cruise Tourism Development as the Starting Points to Make Innovation in the Early and Pilot Implementation System

Firstly, for cruise services, the transit-mode supervision on food of cruise services shall be actively promoted. In August 2015, the Shanghai Entry-Exit Inspection and Quarantine Bureau formulated the *Inspection Management Provisions on Food Supply Chains for Transit Cruise Supply (Trial)*,[16] which changes entry regulation on food for cruise services to transit one, promoting the development of cruise food delivery industry in Shanghai. And it accelerates the construction of Shanghai cruise home ports to open fast track for foreign food's direct supply to cruise ships, and to provide exercisable way for convenient customs clearance. For transit-food for cruise supply, inspection and quarantine departments are required to do a good job in supervision and inspection before the event, at present and after the event in accordance with regulatory requirements on food supply chains for transit cruise supply, so as to protect food safety and health of tourists on the cruise ship. Actively promote tariff-guarantee and customs-transit supply on food for cruise ships, which can provide more convenient and perfect services for international cruise ship while operating at the home port, and build the home port into a material allocation and delivery center for international cruise services with centralized purchase, regulation, delivery, settlement and other functions. Therefore, the cruise industry's economy driving role will be truly played.

Secondly, conduct cruise tourism on high seas and along the coast for early and pilot implementation in FTZ and experimental area for cruise tourism. With incidental business along the coastal at FTZ as a system breakthrough, actively promote the development path of cruise tourism along the coast and cruise international tourism coinciding well with domestic one. Create domestic high-quality routes to attract tourists at home and abroad, thus driving pulling effect at coastal ports. Set Chinese-funded cruise lines as a pilot projects, and actively promote cruise tourism along the coast to form flexible demonstration effect of the foreign-funded cruise lines, and actively try routes for cruise tourism on high seas in the form of Memorandum of Understanding signed by Ministry of Public Security. As South Korea is affected by Sade event this year, the cruise lines have to adjust their routes with ports in South Korea for the Japanese ones. Among them, part of the four-day-and-three-night and one-way voyage to Cheju has been changed to the so-called route of "First Class at Sea Experience Tour" without a destination. Set the existing mode as a sample, and actively explore the routes without a destination.

---

[16]Lei Yang, Discussion on Diversified Business Development of Bonded Warehouse at Shanghai Port International Passenger Transport Center, *Contemporary Economics*, 2016/Issue 20: 28.

Thirdly, for the dilemma of departure tax refund for cruise ships, we can actively learn from the practice of duty-free shops at airports to set up entry duty-free shops at cruise ports so as to avoid the outflow of consumption and enhance the supporting business environment at cruise home ports of cruise industry, giving full play to huge driving effect of cruise industry economy.

Acknowledgments This article is funded by the following projects: Investigation and Research on Foreign-related Disputes over Maritime Fishing o f Liaoning Province under the Strategy of "the Belt and Road" 2017lslktjd-012 and Research on Maritime Crime Prevention Measures of the Province 2017lslktjd-014.

Chapter 11
Discussion on the Nature and Governance Mode of Cruise Ship Environmental Pollution from the Perspective of Marine Right

Ruihong Sun, Hong Wang and Xinde Chen

1 Introduction

The 21st century is a marine economy era (Xi 2017). As the golden industry of golden waterway, cruise tourism plays a vital role in international tourism and transnational trade and it also serves as a critical part of China's marine economy. Developed since 2006, China's home port cruise tourism has become the cruise ship market with the fastest development speed around the whole world (CLIA 2016). However, as the huge ships used for both transportation and tourism, they are featured by "high energy consumption, high emission and high pollution" and have become the main source for the coastal and marine environmental pollution. According to the policies promulgated by U.S. Environmental Protection Agency, "all cruise ship companies shall use low–sulfur fuel since 2015". The report predicted that the measure would avoid the premature death of 12,000–31,000 persons per year and save medical expenditure of USD 110–270 billion by 2030 (EPA 2008).

Fund project: National Geographic Air and Water Conservation Fund (GEFC24-14); general project of National Social Science Foundation (16BGL110), humanities and social sciences project of Ministry of Education (13YJC630207); Shanghai Municipal People's Government Development Research Center—Shanghai University of Engineering Science "government public decision-making support" study base special case (2017-YJ-B15), Tourism Young Expert Training Program of China National Tourism Administration (TYEPT201416, TYEPT201523).

Ruihong Sun—Research orientation: Cruise ship tourism and Environment management

Hong Wang—Research orientation: Cruise ship economy and Social security

Xinde Chen—Research direction: Cruise ship economy

R. Sun (✉) · X. Chen
Shanghai University of Engineering Science, Shanghai, China
e-mail: ruihongsun@126.com

H. Wang
Shanghai International Cruise Business Institute, Shanghai, China

© Social Sciences Academic Press and Springer Nature Singapore Pte Ltd. 2018
H. Wang (ed.), *Report on China's Cruise Industry*,
https://doi.org/10.1007/978-981-10-8165-1_11

The marine environment pressure and pollution brought by the fast development of cruise ship consumption and production greatly challenge the safeguarding of China's marine rights and interests and sustainable development of marine economy. In consideration of the later development of China's cruise ships, incomplete policies for cruise ship pollution prevention and weak supervision on cruise ship pollution, environment governance needs to be enhanced urgently.

2 Research Review

2.1 Research Abroad

The research of foreign academic circle on cruise ship environmental pollution could be traced back to Erize (1987), who once predicted that the cruise ships, which were mainly used for an increasing number of tourists to travel to the South Pole, would exert adverse influence on the ecological environment of the South Pole; he also suggested protecting the environment of the South Pole by controlling the number of tourists and strengthening education. Later, Ritter and Schafer (1998) claimed that tourism study had paid much attention to the sustainable development of offshore tourism in the long run but did not highlight the environmental pollution brought by the cruise tourism of sea, which accounted for 2/3 area of the whole earth. He believed that the main reason why cruise ship pollution was neglected was that cruise ship was an emerging travel mode and accounted for a low proportion in international tourism. However, with the increase in cruise tourism, the environmental pollution it triggered would become considerable. With the popularity of cruise tourism, more researches were made by the academic circle since 2000 along with more attention paid to cruise ship pollution. Most researches were focused on the environmental pollution caused by illegal dumping of cruise ships, oil dirt emission and stranding (Podolsky 2004; Rogers and Garrison 2001); pollution supervision and governance (Morehouse and Koch 2003) were proposed and attention was paid to the sustainable development of cruise ship economy (Johnson 2002) in these researches, where discussion was also made on the applicable laws and regulations and governance of cruise ship pollution (Copeland 2008; Morehouse and Koch 2003). Generally speaking, cruise ship pollution in this period did not attract enough attention of the academic circle and the research results were limited.

Researches on cruise ship pollution gradually became active and they could be divided into three fields from the previous odd researches since 2010: the first was to discuss the environmental impact of cruise tourism pollution from the ecological perspective by combining different regions, such as coastal port (Carić 2011), waters, (Carić and Mackelworth 2014; Dawson et al. 2014), fjord (Mölders et al. 2013), island (Cheer 2016), coral reef and ecological zone (Diedrich 2010) and to evaluate the economic and social cost of pollution (Carić 2011; Maragkogianni and Papaefthimiou 2015); the second was to highlight atmospheric pollution and carbon emission of cruise ships; carbon emission of cruise ships, routes and tourism were evaluated according to quantitative method (Howitt et al. 2010) and the influence of carbon emission on the environment was discussed (Eckhardt et al. 2013; Poplawski et al. 2011); the third was that the sustainable development of cruise tourism (Genç 2016) included cruise pollution assessment (Carić 2016a), sustainable and flexible marine planning (Adams 2010), cruise ship control framework and implementation, improvement of applicable policies and laws (Copeland 2011), environmental behaviors of cruise ship companies and international cooperation in the governance of transnational pollution (Morehouse and Koch 2003). During the period, cruise ship pollution works published by Miller et al. (2013) analyzed the nature of cruise ships and pollution types, the influence of pollution on economy, society and environment as well as the cruise ship pollution supervision and governance comprehensively.

2.2 Research at Home

The first person who proposed cruise ship pollution in domestic academic circle is Qiu (2007), who brought forth suggestions on the environmental pollution that might be brought by cruise ship economy from the perspective of prevention. The early domestic research on cruise ship pollution was mainly about the historical experience of foreign cruise ship pollution prevention and international laws and policies on cruise ship pollution prevention. In this period when the domestic home port cruise tourism just started and cruise ship pollution had not appeared yet, there are only a few domestic researches made. After 2010, domestic scholars paid more attention to and put more efforts to the research of cruise ship pollution, due to the occurrence of two events: the first one is that China committed in 2009 that by 2020, the unit GDP carbon dioxide emission would be reduced by 40–45% than that in 2005; the second is about the stranding and grounding of Costa Concordia Cruise Ship. Therefore, it can be concluded that domestic research on cruise ship pollution is mainly focused on two parts: the first one is about the calculation of cruise ship carbon emission and ecological footprint (Sun et al. 2014), the development of shore power technology (Wu and Shen 2012) and the construction of cruise ship port power (Du et al. 2016); the second is about the reference to the foreign experience in cruise ship pollution governance, such as cruise ship specifications (Sun et al. 2015, 2017), low-carbon design shore power experience (Du et al. 2016), discussion on the prevention of Chinese cruise ship pollution (Wang and Huang 2015), calculation and control of cruise tourism environmental pollution (Li and Lv 2016) and ecological monitoring and restoration.

3 Thinking on the Nature of Cruise Ship Environmental Pollution

3.1 Cruise Ship Pollution Being Typical and Representative in Marine Pollution

Cruise ships, regardless of its low proportion in the whole marine ships, consume more energies and substances and discharge more wastes than other types of ships, due to its large tonnage, multiple functional facilities and high passenger capacity (the maximum passenger capacity of Harmony of the Seas is 6410 persons). Cruise ships travel along the same route in the long run, exerting significant influence on the local environment (Copeland 2008). By taking sulfide emission as an example, the sulfide emission of heavy fuel oil generally used by a cruise ship is millions times higher than that of a car when traveling for the equivalent distance, which results in serious air pollution and health threat.

Cruise ship pollutants include waste water (black water, grey water, including oily bilge water and ballast water), solid waste (sludge, general solid waste and hazardous solid waste) and waste gas (incinerator and engine emission), all of which exert significant influence on the environment (Miller et al. 2013); as shown in Table 1, cruise ship pollution deteriorates marine ecological environment, destroys the critical habitat of marine organisms (Diedrich 2010), reduces the quality of marine tourism resources and even causes damages to the physical health of the coastal citizens (Carić 2016b); it also increases ecological and social costs (Maragkogianni and Papaefthimiou 2015). With the fast development of cruise ship economy and more attention paid to ecology, the negative influence caused by cruise ships on sea and coastal ecological environment becomes prominent gradually and has attracted more and more attention.

3.2 Increase in Cruise Ship Pollutants Becoming Faster and the Caused Loss Becoming More Prominent

Cruise ship environmental pollution increases rapidly and the caused loss to economic development becomes increasingly prominent, so environmental governance becomes an urgent task. At present, there are two methods for cruise ship pollution and its environmental impact assessment: the first one is carbon emission calculation and carbon footprint evaluation (Eckhardt et al. 2013; Howitt et al. 2010; Mölders et al. 2013; Poplawski et al. 2011); the second is calculation of the specific pollutants and evaluation on its environmental impact by ecological indicators (Butt 2007). According to the current researches, both methods have revealed the positive correlation of cruise ship pollution and cruise ship development. Figures 1 and 2 are the method for calculating pollutants used by the research group according to the existing document research, both at home and abroad; the carbon emission and

Table 1 Cruise ship environmental pollution and its impacts

Pollutants	Type of pollutants	Environmental pollution	Environmental impact
Waste water	Black water	Polluting water quality and infecting bacteria and virus	Not beneficial to fishing industry
	Grey water	Water eutrophication and increase in marine plants	Oxygen deficit of water environment and death of fishes
	Oily bilge water	Shading sunshine and oxygen integration and destroying sea photosynthesis	Destroying the critical habitat of marine organisms and threatening marine organisms and birds
	Ballast water	Invasion of alien species	Destroying the integrity of ecological system and doing harm to the growth and yield of local organisms
Solid wastes	Sludge	Seawater eutrophication and oxygen deficit of water environment	Oxygen deficit of water environment and death of fishes
	General solid wastes	Polluting water quality and blocking channels	Doing harm to navigation and fishing industry
	Hazardous solid wastes	Hardly degraded naturally with long-term toxicity	Infecting marine organisms
Waste gas	Incinerator	Emitting heavy metal and harmful substances	Damaging the health of living bodies
	Engine emission	Emitting oxynitride, oxysulfide, diesel particulate matters and carbon dioxide	Causing acid rain, affecting visibility and doing harm to the respiration system of both humans and animals

pollutant growth along with the development of China's cruise tourism are analyzed from the two perspectives above; data show that China's cruise ship pollution and the pollution loss caused therefrom are on the rise very quickly over the past two years, so pollution governance becomes an urgent task.

3.3 *Cruise Ship Pollution Being Transnational and Global*

In light of the fact that cruise ship pollution is quite special, the emission of cruise ship pollutants is transnational and global due to the transnational features of cruise ship companies, the international features of routes and the global mobility of marine ecological system and the pollution will affect different countries and even the whole world along with the mobility of sea and atmospheric environment. Transitional cruise ship pollution has three connotations: the first is the environmental pollution

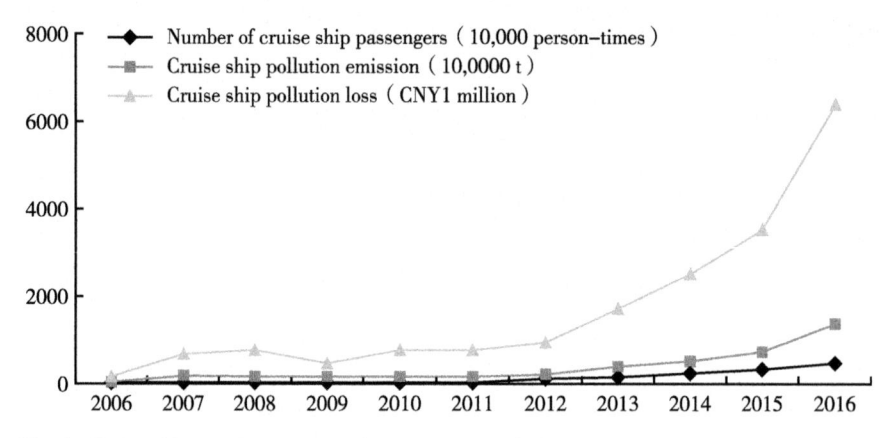

Fig. 1 Cruise ship population and pollution growth in 2006–2016

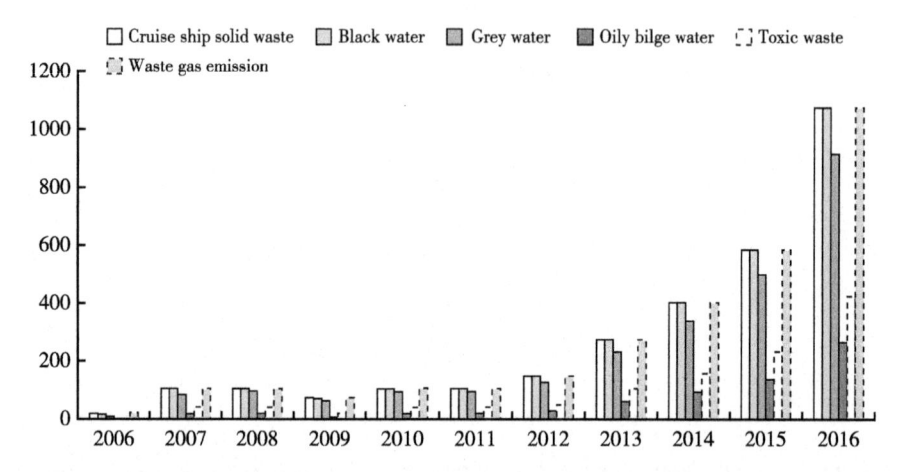

Fig. 2 Cruise ship pollutant growth in 2006–2016

of transnational cruise ship companies; the second is the cruise ship pollution happened in one country but affecting the environment of different countries; the third is that transnational cruise ship route brings environmental pollution to other countries and even to the whole world. "Tragedy of the commons" of marine environmental pollution caused by "transnational" cruise ship pollution makes cruise ship pollution governance more difficult; first, transnational governance involves the interests of different countries and different parties, including government, port community residents, cruise ship companies, etc. (Zhao and Bai 2012); once effective public waste management is not available, more floating marine garbage will be caused; secondly, cruise ship transnational pollution governance needs to be supported by international laws, bilateral or multilateral agreements and the serial coordination of local laws to ensure its effective execution. At present, international cooperation in cruise ship pollution governance needs be reinforced (Yin 2016); thirdly, the transnational and mobile features of cruise ship operation make cruise ship pollution uneasy to be found and the pollution of marine environment caused can be easily ignored (Carić 2016a); a majority of countries, among which, cruise ship travels, do not have effective pollution control mechanism and effective environmental monitoring and management system (Maragkogianni and Papaefthimiou 2015); fourthly, it is greatly difficult to acquire relevant techniques and knowledge in cruise ship pollution governance (Ballini and Bozzo 2015) and the international technical cooperation among different countries needs to be strengthened.

3.4 Marine Right Features Involved in Cruise Ship Pollution Governance

In light of the transnational operation of cruise ship, international subject and global mobile marine environment, the transnational cruise ship environmental pollution government should be considered from the perspective of safeguarding the maritime sovereignty interests. As explicitly specified in the report of the 18th National Congress of the Communist Party of China, "We should protect ecological environment, safeguard national marine rights and interests firmly and construct a maritime power". This provides critical orientation and clearly defines the task for the discussion of transitional cruise ship pollution governance from the perspective of marine right. The president Xi Jinping proposed in Davos World Economic Forum held on January 17, 2017 that we should advocate active multilateral cooperation to cope with climate change jointly and construct a new pattern of relationship between great powers; it means that China has got prepared for playing a more important role in the world and shouldering more responsibilities and highlights China's management right in international relationship and the leadership authority in the global environment governance. It also means marine right will play a more important role. Therefore, to investigate the transnational mobility of cruise ship pollution from the perspective of marine right and think about the bottleneck of transnational cruise ship pollution and China's pollution governance by combining the current domestic pollution prevention laws and applicable international regulations on marine pollution is beneficial to the proposal of the cruise ship pollution governance mode that best fits China.

4 Challenge and Mode Discussion of Cruise Ship Pollution Governance

4.1 Governance Challenges Brought by Cruise Ship Pollution

Cruise ship pollution brings great challenges to China's cruise ship pollution governance, but the following bottlenecks also exist in consideration of China's transnational pollution governance ability and realities: Firstly, China should show more undertakings and has more right of speech as a great power in the global environmental governance and get rid of the difficulties that China had to participate in transnational pollution governance passively (Jiang 2014). Secondly, China needs to establish a corresponding guarantee system; the waste treatment industrial system centering on cruise ship has not been formed, the protection of port environment by policies is insufficient, a complete marine ecological compensation mechanism has not been established and effective cruise ship pollution prevention policies have not been proposed (Wang and Huang 2015). Thirdly, China needs to strengthen its researches on cruise ship pollution, establish the emission inventory of port atmospheric pollutants, assess the cost required by pollution control measures and benefits that it may bring and try to establish emission control zone (Li and Lv 2016). Fourthly, we should coordinate management mechanism, promote the smooth implementation of environment governance through the supporting administrative system reform and rationalize marine environment management system.

The internationality of cruise tourism and the globalization, penetration, nonopposability, nonexcludability and externality of marine ecological environment determine that cruise ship pollution governance entails the involvement of all government subjects and its multilateral cooperation governance mechanism should be analyzed from the perspective of global governance. Therefore, the decision-making process for cruise ship pollution governance in future needs the participation of more interest groups; it is also of great importance to promote cooperative environment governance and lay more emphasis on the multilateral cooperation during the process of treating social matters (Gareau 2015) to coordinate the interests of all parties with the lowest cost and reduce interest conflicts of all parties (Fan and Yin 2016). In addition, environment governance reformers and managers should determine the environment governance mode that is in line with the idea of "actively advocating multilateral cooperation, coping with climate change jointly and constructing a new pattern of relationship between great powers" on the basis of China's demand for marine right and great power status and seek for the integration of environment governance mode under Chinese context so as to realize tolerant environment governance.

4.2 Thinking on the Cruise Ship Pollution Governance Mode

It is required to deeply understand the international cruise tourism environmental pollution, analyze the evolvement process of cruise tourism pollution and governance by using historical and literature analysis, the economic and ecological problems brought by transnational cruise ship pollution as well as externalities of cruise ship environment and social costs, predict the pollution trend and judge its toughness; the bottleneck of transnational cruise ship pollution governance and orientation should be defined by combining China's current realities of cruise ship pollution governance and by referring to the foreign transnational experience and examples in pollution governance. Generally speaking, China needs to consider the transnational pollution and multiple subjects involved in cruise ship pollution governance from the following several aspects:

Firstly, multi-agent game and cooperation mechanism in transnational cruise ship governance should be revealed. Interest analysis method can be used to recognize relevant interested parties of transnational cruise ship pollution, such as sovereignty, intergovernmental organizations and non-government organizations, so as to establish multi-layer 3D framework for transnational cruise ship pollution governance from multi-dimensional perspective; the correlation and mechanism of action of multiple subjects in transnational cruise ship governance as well as the complex structure features of the joint efforts of multiple subjects in this field should be studied and judged from the perspective of global environment governance; the selection of strategies adopted by different multiple subjects against transnational pollution governance is analyzed by establishing transnational cruise ship pollution governance model involving surrounding countries and multiple subjects; the governance model participated in by multiple governance subjects is analyzed to discuss the closed cooperation of all parties in cruise ship pollution governance and construct cooperation and win-win based multilateral governance mechanism so as to realize perfect Nash equilibrium.

Secondly, multilateral cruise ship pollution governance mode dominated by the marine right of China should be constructed. Multilateral environmental governance structure under modern marine right should be designed by combining the current international order transformation and global governance revolution, by reviewing our marine right claim in transnational cruise ship pollution governance on the basis of utility decision-making analysis and by considering the marine rights and interests of all surrounding countries and all interested parties: the division of rights and responsibilities of multiple subjects in cruise ship pollution governance should be investigated and analyzed from the perspective of rights and responsibility counterbalance and symmetry on the principle that "he who pollutes it shall govern it" and co-sharing with different responsibilities so as to realize the balance and unity of marine ecological environmental protection and pollution governance responsibilities, marine resources environment control and power utilization, marine politics, economy, ecology and cultural interests.

Thirdly, the realization path and guarantee system of multilateral cruise ship pollution governance should be implemented. International laws and international relationship criteria and conventions should be improved and the corresponding domestic policies and laws should be promulgated; the example set by great power in multilateral governance and multidirectional public participation should be ensured by legal systems to promote the effective cooperation of cruise ship pollution governance; stable, long-lasting and effective capital input operation mechanism, efficient international technical transfer mechanism and public information platform should be explored to develop public guarantee for the development of multilateral environment governance, such as capital, technique and information; multilateral cruise ship governance committee should be established to coordinate the interests of different parties, negotiate on conflicts, manage public matters, take combined actions and provide personnel and organization operation guarantee system for cruise ship pollution governance.

5 Conclusion and Prospect

The existing researches at home and abroad are mainly on the source, calculation, occurrence mechanism and environmental impact mechanism of cruise ship pollution, but less on the transnational cruise ship pollution and its environment governance. By starting from cruise ship pollution, the cruise ship pollution governance mode is studied in this paper and the cruise ship environment governance mode is discussed through the overview on domestic and foreign researches on cruise ship pollution and analysis on pollution nature in combination with the realities of Chinese cruise ship development. The following points of view have been formed in this paper:

Research on cruise ship pollution should consider the transnational pollution of cruise ship companies and global mobility of cruise ship pollution; the development of cruise ship pollution governance from the perspective of global governance should be discussed. The internationality and ecological globalization of cruise ship make cruise ship pollution governed by different countries or even through the efforts of the globe; the bottleneck of cruise ship pollution governance lies in breaking the value dilemma (marine right claim), cooperation dilemma (multi-agent game) and system dilemma (systems and mechanisms) in multilateral governance. The research on cruise ship pollution governance is enhanced from focusing on the sustainable development strategy of cruise tourism and pollution control and prevention to the comprehensive cruise ship environmental pollution governance. Therefore, not only the multilateral international relationship and interest game but also legal and technical cooperation should be considered; effective research integrating methods of different disciplines are urgently needed.

Transnational cruise ship pollution governance involves multilateral subjects and how to strike balance of nation rights and interests and multilateral interests becomes more important than ever. For the purpose above, benefit game of multilateral subjects and rent in the transnational cruise ship pollution governance should be resolved urgently; marine right theory is introduced by combining the enhancement of China's comprehensive national strength, change in relationship between major countries and enhancement of global governance status along the change in new type international relations to review China's role in global governance and to construct the multilateral governance mode under China's subject viewpoint.

Traditional environment governance mode is determined under the dominance of hegemonic power under international relations at that time; with the fast rise, China, as a responsible great power and active participant in global governance, needs to reposition the benefits of maritime sovereignty and build the marine right idea of "environmental justice, cooperation and win-win". Cruise ship pollution governance needs to be designed according to the specific responsibility and right system, by representing the environmental justice principle that "he who pollutes it shall govern it" and the responsibility shouldering principle of "co-sharing with different responsibilities" for developing countries, besides highlighting efficiency and fairness. At the same time, pollution governance also entails domestic and international systems and mechanisms and public guarantee; in this process, China is also required to enhance the corresponding governance capacity to realize cruise ship pollution governance.

Chapter 12
Analysis on the Development Strategy and Path of China's Domestic Cruise from the Perspective of Differentiated Competition

Xinliang Ye, Junqing Mei and Yuanqin He

1 Introduction

As the birthplace of global cruise tourism, North America is the world's largest traveler generating region and tourist destination region for cruise tourism, accounting for nearly 60% of the global market share. Miami, Florida is also called the "cruise capital" in the world. With the gradual saturation of the European & American market and the transfer of international cruise companies' strategic objectives, the Asia-Pacific region, as one of the most dynamic regions of the world economy, transfers the focus of the cruise fleet to the Asia-Pacific region, creating the eastward shift of the international cruise industry, which brings a major opportunity for the development of China's domestic cruise enterprises. According to the forecast of Cruise Lines International Association, the global cruise market is expected to reach 25.8 million people in 2017, 26.9 million in 2018 and

---

Xinliang Ye, male, born in Zhenhai, Zhejiang; Research direction: Cruise tourism and Risk management.

Junqing Mei, male, born in Kaifeng, Henan; Research direction: Cruise tourism.

Yuanqin He, female, born in Lu'an, Anhui; Research direction: Cruise tourism.

Fund Project: General project supported by National Social Science Fund (16BGL110), project supported by Ministry of Education Humanities and Social Science (13YJC630207); Tourism Youth Expert Training Program of China National Tourism Administration (TYEPT201416, TYEPT201523).

X. Ye (✉)
Shanghai University of Engineering Science, Shanghai, China
e-mail: yexinliang@sues.edu.cn

X. Ye · J. Mei
Shanghai Wusongkou International Cruise Terminal, Shanghai, China

X. Ye · Y. He
Shanghai International Cruise Business Institute, Shanghai, China

© Social Sciences Academic Press and Springer Nature Singapore Pte Ltd. 2018
H. Wang (ed.), *Report on China's Cruise Industry*,
https://doi.org/10.1007/978-981-10-8165-1_12

37.6 million in 2025, indicating that the international cruise market still has good prospects for development and huge market potential.

According to Fig. 1, the growth rate of the international cruise market is significantly higher than the development speed of the global economy. Under the global financial crisis in 2008, the international cruise market still maintained a growth rate of 4%. In 2009, facing with the sharp fall of global economy, it kept a growth rate of 7.9%. However, it can also be seen that the growth of the international cruise market has a high correlation with the development of the global economy, which has a direct impact on the disposable income level of the residents. The development of the global economy is influenced by many factors, such as population change, scientific and technological progress, domestic politics, production flow, international rules and international system. By contrast, the growth of tourism economy is mainly affected by the growth rate of tourism investment, tourism resource supply, political and economic situation, tourism consumption demand and psychological expectations of tourists, which have the most direct relationship with the growth of the economy (Fig. 2).

Asia-Pacific cruise market is mainly divided into three regions: the South Pacific region represented by Australia, New Zealand and Indonesia, the Southeast Asia region represented by Malaysia, the Philippines, Singapore, and India, and the Far East region represented by China, Japan, South Korea and North Korea. In the past decade, the Chinese cruise market has been well nurtured. In 2016, China became the largest cruise market in the Asia-Pacific region, with the number of cruise tourists reaching 2.28 million. As the most dynamic region of the Asia-Pacific cruise market, China is gradually being developed into the new center of the global cruise market, providing a good market base for the development of China's domestic cruise brand.

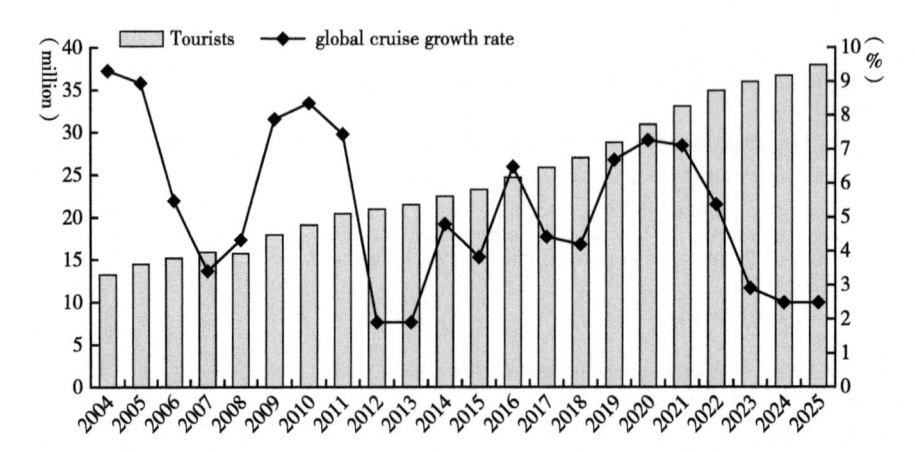

Fig. 1 Changes in scale and development trend of global cruise market. *Source* Cruise Lines International Association

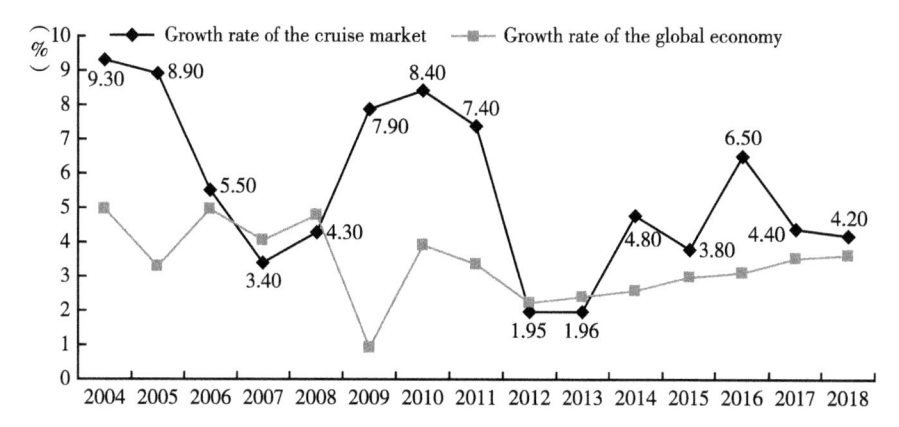

Fig. 2 Changes in development trend of global cruise market and global economy. *Source* Collected by the author

2 Analysis on the Development Environment of Domestic Cruise Market

2.1 Asian Cruise Market Has Become an Important Emerging Market in the World

From the perspective of the transport capacity of the international cruise market, Asian cruise capacity accounts for 9.2% of the global cruise market, ranking the fourth in the world; the Caribbean region ranks the first with a ratio of 33.7%, with a year-on-year growth of 2.8%; the Mediterranean region ranks the second with a ratio of 18.7%, with a year-on-year growth of 2.0%; the European region ranks the third with a ratio of 11.6%, with a year-on-year growth of 8.7%; Australia/New Zealand/Pacific region ranks the fifth with a ratio of 6.1%, with a year-on-year growth of 19.2%; the Alaska region ranks the sixth with a ratio of 4.1%, with a year-on-year growth of 1.7%; the South America region ranks the seventh with a ratio of 2.8%, with a year-on-year growth of 5.3%; other regions account for 13.8%, with a year-on-year growth of −1.4%. In 2016, the Asian cruise capacity accounted for 9.2% of the global cruise market, with a year-on-year growth of 38% over 2015 and 193.5% over 2012. According to the growth rate of cruise capacity in recent years, 2013 saw an increase of 55.6% compared with 2012, 2014 saw 16.6% over 2013 and 2015 saw 17.7% over 2014, indicating that the Asian region is the world's fastest growing cruise market. With respect to the cruise capacity, the number of Asian cruise brands reached 31 in 2016, which was 26 in 2015, the number of cruise ships was 60, the number of voyages reached 1560, the total berthing volume reached 5500, the total operating days of the ports reached 7918, the number of cruise destinations reached 204, and the potential number of the visitors' staying at the destinations reached 10.9 million (Fig. 3).

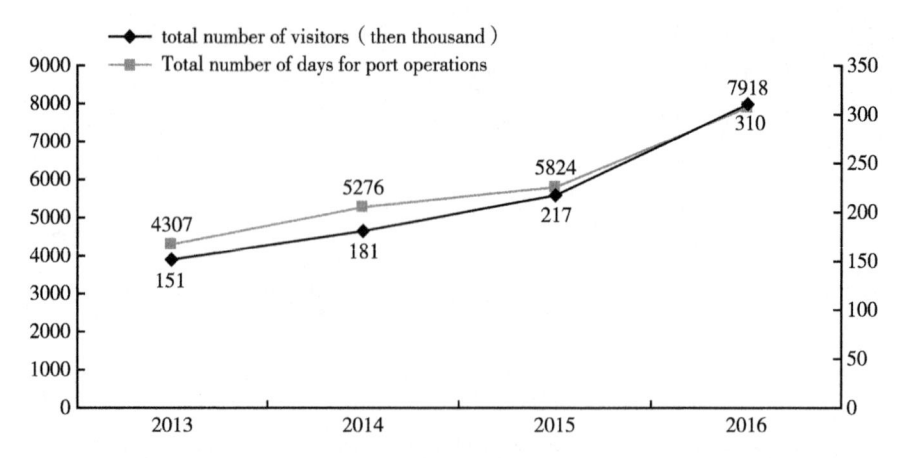

Fig. 3 Changes in total number of days of cruise operations in Asia and that of tourists in recent years. *Source* Cruise Lines International Association

In terms of the supply of Asian cruise products, among the 1560 voyages, the departure and arrival regions of 1473 voyages are in Asia, and this figure was 803 in 2013, while the number of other international voyages was 87. The vast majority of Asian cruise routes are based on the Asian region, and a rapid growth is shown. Due to features of 48-h sailing circle and travel by night and rest by day, the Asian region is the best sailing region, which allows the visitors to fully experience the leisure of cruise tourism and the cost is acceptable.

According to the capacity of different cruise companies, the capacity of the cruise ships under Star Cruises is declining. The main internal factor is that the new ships are less and smaller, and the external factor is that the cruise companies in the Asian cruise market are increasing gradually. Besides, few efforts have been made by Star Cruises in the Chinese cruise market, while the Royal Caribbean and Costa Cruises are making great efforts in entering into the Chinese market by investing larger and newer cruise ships, which significantly enhances their capacity. Costa Cruises entered into China in an earlier time. With the gradual increase of the capacity layout, Costa Cruises occupies the largest market share in Asia, with its passengers mainly come from China. The Royal Caribbean leads China enter into the great cruise era. Benefiting from its good brand marketing and good relations with the travel agency, the Royal Caribbean saw a rapid increase in market attractiveness, a gradual enhancement of market competitiveness and a fast expansion of market share (Fig. 4).

According to the types of the 60 ships invested by the international cruise companies in 2016, 21 ones are medium-sized cruise ships, 15 ones are large cruise ships, 16 ones are small cruise ships, 6 ones are adventure cruise ships and 2 ones are giant cruise ships. From this point of view, the proportion of giant cruise ship is still very low, which will lead to a substantial increase in the crowding degree, operating costs and operational difficulties. The large and medium-sized cruise ships account for 60%. There is an obvious trend of increase in the large-sized

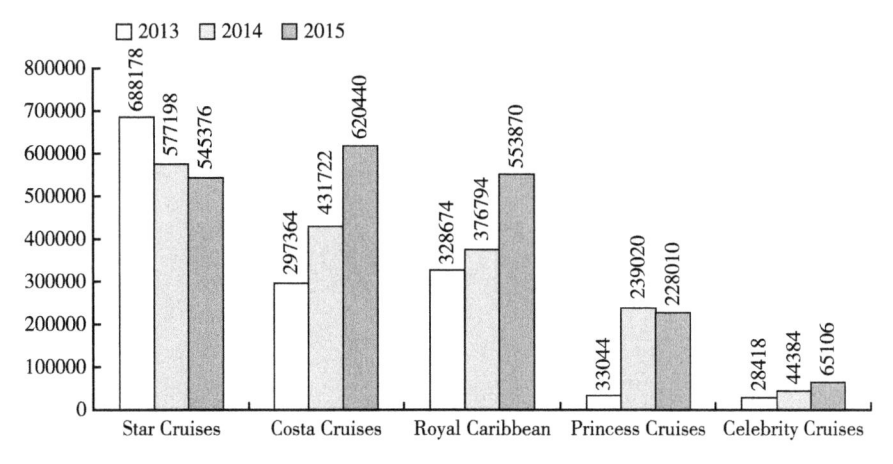

Fig. 4 Changes in capacity of Asian cruise brands in recent years. *Source* Cruise Lines International Association

cruise ship, which may be attributable to the popularity of the cruise tourism. The large-sized cruise ship can meet the diverse needs of tourists and may be equipped with more restaurants and entertainment facilities to create a spacious space for tourists and improve the benefits, highlighting the nature of cruise ship as the destination, modern city over the sea. However, the crowded cruise ship will have a certain impact on the cruise experience.

2.2 The Rapid Growth of Chinese Cruise Market Has Provided a Favorable Environment for Domestic Cruise Brands

According to the statistics of China Cruise & Yacht Industry Association (CCYIA), China's 11 largest cruise terminals (Dalian, Tianjin, Qingdao, Yantai, Shanghai, Zhoushan, Xiamen, Shenzhen, Guangzhou, Haikou and Sanya) received 1010 cruise ships in 2016, with a year-on-year growth of 61%, and the number of visitors reached 4.567 million, with a year-on-year growth of 84%. Among them, China received a total of 927 cruise ships from home ports, with a year-on-year growth of 72%, and 83 ones from visit ports, with a year-on-year growth of −8%; the number of inbound and outbound tourists from home ports reached 4.2897 million, with a year-on-year growth of 93%; the number of inbound and outbound tourists from visit ports reached 277,500, with a year-on-year growth of 8%. Shanghai Cruise Terminal received 509 international cruise ships, with a year-on-year growth of 48%, and the number of inbound and outbound tourists reached 2.944 million, with a year-on-year growth of 79%, ranking the first among national cruise ports (Table 1).

Table 1 Cruise ships and tourists received in China's cruise terminals in 2016

Terminal	Home port Voyages	Home port Tourists	Visit Voyages	Visit Port Tourists	Total voyages 2016	Total voyages Annual growth	Total tourists 2016	Total tourists Annual growth	Ranking
Shanghai	481	285.9	28	8.56	509	48%	294.4	79%	1
Tianjin	128	68	14	3.51	142	53%	71.56	66%	2
Guangzhou	104	32.6	0	0	104	10,300%	32.59	12,437%	3
Xiamen	66	14.37	13	5.71	79	20%	20.8	14%	4
Qingdao	52	8.6	0	0	52	174%	8.6	168%	5
Haikou	39	6.18	2	0.27	41	58%	6.45	685%	6
Dalian	27	6.48	0	0	27	35%	6.48	183%	7
Sanya	0	0	25	9.64	25	−17%	9.64	7%	8
Shenzhen	14	4.45	0	0	14	–	4.45	–	9
Zhoushan	12	1.73	1	0.0405	13	8%	1.77	−11%	10
Yantai	4	0.57	0	0	4	−78%	0.5784	−49%	11
Total	927	428.88	83	27.7305	1010	–	457.3184	–	–

Source China Cruise & Yacht Industry Association (CCYIA)

Table 2 List of home-port cruise ships in China in recent years

No.	Home-port cruise ships in China in 2015	No.	Home-port cruise ships in China in 2016	No.	Home-port cruise ships in China in 2017
1	Quantum of the Seas	1	Quantum of the Seas	1	Quantum of the Seas
2	Mariner of the Seas	2	Ovation of the Seas	2	Ovation of the Seas
3	Voyager of the Seas	3	Voyager of the Seas	3	Voyager of the Seas
4	Legend of the Seas	4	Legend of the Seas	4	Mariner of the Seas
5	Victoria	5	Mariner of the Seas	5	Serena
6	Serena	6	Serena	6	Atlantic
7	Atlantic	7	Atlantic	7	Victoria
8	Sapphire	8	Victoria	8	Lucky
9	Superstar Libra	9	Lucky	9	Poesia
10	New Century	10	Poesia	10	New Century
11	Henna	11	New Century	11	Brilliant of the Seas
12	China Taishan	12	Brilliant of the Seas	12	China Taishan
		13	China Taishan	13	Sapphire
		14	Sapphire	14	Majestic
		15	Golden	15	Superstar Virgo
		16	Superstar Libra	16	Genting Dream
		17	Superstar Virgo	17	Joy
		18	Genting Dream	18	Silver Shadow

In 2017, China has 18 home-port cruise ships. Among them, 4 ones for the Royal Caribbean, namely Quantum of the Seas, Ovation of the Seas, Voyager of the Seas and Mariner of the Seas; 4 ones for Costa Cruises, namely Serena, Atlantic, Victoria and Lucky; 1 for MSC Cruises, namely Poesia; 1 for SkySea Cruise Line, namely New Century; 1 for Diamond Cruises, namely Brilliant of the Seas; 1 for Bohai Ferry, namely China Taishan; 2 ones for Princess Cruises, namely Sapphire and Majestic; 1 for Star Cruises, namely Superstar Virgo; 1 for Dream Cruises, namely Genting Dream; 1 for Norwegian Cruise Line, namely Joy; 1 for Silversea Cruises, namely Silver Shadow. Although the entry of international cruise brands into the Chinese market brings great competitive pressure to the domestic brand, it has an important role in expanding the China's cruise tourism market and enhancing its cruise culture (Table 2).

2.3 *China's More Support to the Development of Domestic Cruise Ships*

With the entry of international cruise companies into the Chinese market, "domestic cruise", "domestic cruise business", "domestic cruise tourism", "China customized cruise", "offshore cruise tourism" and other words gradually become the hot topics in the cruise sector. In recent years, the North American market has gradually changed from the "superstar" to the "Taurus" in the Boston Matrix. The international cruise companies need to seek emerging markets to maintain the high growth rate of the cruise market, and China has become their best choice. The facts have proved that this choice is correct. The growth rate of Chinese cruise market is above 40%, far more than expected. However, Chinese cruise market is facing with many problems with the international cruise company plays a dominant role in the Chinese market, for example, the Chinese cruise ship is less competitive, the cruise tourism has a low contribution to the regional economy, and the growth rate of outbound tourists is much higher than that of the inbound tourists, the trade deficit is becoming more and more serious. The Chinese government is urged to enhance China's initiative in the international cruise market, improve our economic contribution to the international cruise market, establish the international cruise brand and increase supports for the development of domestic cruise brand (Table 3).

Table 3 Relevant policies supporting the development of domestic cruise brands

National level	Policy contents
Made in China 2025	Make breakthrough in key areas to develop techniques for the design and construction of luxury cruise ships
Notice of the State Council on Printing and Distributing the "13th Five-Year" Tourism Development Plan	Encourage multiple capitals to enter the cruise tourism industry to strengthen cooperation with foreign-funded cruise enterprises and support the development of domestic cruise business
Several Opinions of the State Council on Promoting the Reform and Development of the Tourism Industry	Continue to support the localization of cruise ships, yachts and other tourism equipment, and actively develop cruise & yacht tourism
Several Opinions of the State Council on Promoting the Healthy Development of the Maritime Industry	Orderly develop dry bulk cargo transport fleet and cruise economy, consolidate the international competitive position of dry bulk cargo transport and foster regional cruise transport brand
Several Opinions of the General Office of the State Council on Further Promoting Tourism Investment and Consumption	Support the establishment of self-supporting system for R&D, design and construction of large domestic cruise ships and encourage qualified domestic shipbuilding enterprises to design and manufacture large and medium-sized cruise ships
Guiding Opinions of the Ministry of Transport on Promoting the Sustained and Healthy Development of China's Cruise Transport Industry	Actively design and develop cruise products with rich contents and specialized themes to create unique cruise routes and destinations
Notice of the National Development and Reform Commission on Printing and	Foster the development of domestic cruise ships and support the jointly manufacturing of

(continued)

Table 3 (continued)

Distributing the Action Plan to Stimulate Transformation and Upgrading Through Consumption	large cruise projects by domestic and foreign shipbuilding enterprises
Notice of the Ministry of Transport on the Implementation of the Registration of Import Tax of "Ship of Flag of Convenience"	Encourage the registration of Chinese-funded "Ship of Flag of Convenience" and sailing with the flag of the People's Republic of China
Action Plan of the Ministry of Transport on Further Promoting the Supply-Side Structural Reform (2017–2020)	Coordinate with the development and implementation of the overall development plan for national cruise tourism industry and research and develop policies for Chinese cruise companies and "Ship of Flag of Convenience" pilot
Views of the Ministry of Industry and Information Technology on Promoting Development of Tourism Equipment Manufacturing Industry	Actively carry out cruise design and construction standards and specifications to form self-developed brands and technical standards as soon as possible
Plan on the Implementation of Shanghai Service Trade Innovation and Development Pilot	Encourage the cooperation of international cruise companies with domestic enterprises to set up international cruise companies
Opinions on the Implementation of the Strategy of Accelerating the Implementation of the National Free Trade Zone in Tianjin	Vigorously develop the cruise & yacht economy, explore the internationally competitive shipping development mechanism and operation mode
"13th Five-Year" Development Plan of the Integrated Transport Industry in Xiamen	Actively cultivate domestic cruise enterprises and introduce international large cruise companies for the development of cruise services with international standards
Notice of the People's Government of Hainan Province on Printing and Distributing Policies and Measures to Accelerate the Development of Cruise & Yacht Industry	Encourage domestic cruise & yacht enterprises to carry out pragmatic cooperation with international cruise & yacht enterprises to achieve mutually beneficial win-win situation by procurement and leasing, cooperative operation, change of registered address and other means. Form the cruise fleet with independent property right to make breakthroughs for the Chinese national cruise brand By 2020

3 Analysis on the Dilemma of China's Domestic Cruise Industry

Beginning in 2006, China's home port cruise economy is mainly dominated by foreign cruise companies. Local regions have few restrictions on the access of foreign cruise ships, and a lot of support have been given to actively encourage the development of the cruise market, facilitating the formation of the cruise fleet by foreign cruise companies in China's regional market, such as Costa Cruises, the Royal Caribbean, thus creating the situation of the Chinese cruise market dominated by foreign cruise companies.

Though known as the "gold industry floating on the golden waterway", the cruise industry has little contribution to the economy of China, and its "gold properties" have not been showcased. In the Chinese cruise market, the foreign cruise company is the most profitable one, while cruise ports, cruise distributors and domestic cruise companies are under much operating pressure. Among the revenue of China's cruise port, the berthing fee is the most important part. The reduction of the berthing volume will seriously affect the revenue of the cruise port. With the rapid development of the market, many Chinese enterprises began to enter the cruise market. HNA Group, Ctrip Group, Bohai Ferry and other enterprises began to expand the cruise market, but all in the form of single-ship operation. China's domestic cruise market is still weak, with low profitability or even a serious loss.

3.1 A Brief Introduction to China's Domestic Cruise Brands

The domestic brand itself is full of national pride. To define the concept of domestic cruise, we must first find out what is "domestic". The Chinese mainland is the core area of China dominated by the Han population and Chinese culture as known by the western world. It is a geographical description. However, in the definition of the "domestic cruise" concept, the "domestic cruise" refers to the cruise ship registered in China and hanging the five-star red flag in a narrow sense. It emphasizes the connotation of Chinese culture and the elements of Chinese culture. It is a kind of cruise tourism rich in local culture and with significant national characteristics. In general, China's domestic cruise brands mainly include "Henna" of HNA Tourism (out of service), "New Century" of SkySea Cruise Line, "China Taishan" of Bohai Ferry, "Splendida" of Diamond Cruises and "Nanhai Dream" of Nanhai Cruises. Many tourists who have travelled by "Nanhai Dream" are more willing to think it as "classic cruise" (Table 4).

Known as China's first luxury cruise ship, "Henna" is the first luxury cruise ship operated and managed by China. It is purchased by HNA Group from the American Carnival Cruise Group, marking the official entry of Chinese national brand into the cruise tourism market. "Henna" was modified and fitted with the Chinese logo and Chinese description. As a local luxury cruise company in China, SkySea Cruise Line is co-founded by Ctrip and Royal Caribbean Cruises Ltd. The "New Century" cruise, a modified version of the "Celebrity Century", is managed by the Royal Caribbean with the introduction of the Asian chef team. "China Taishan" developed by Bohai Cruise Co., Ltd. under Bohai Ferry Co., Ltd. is known as China's first wholly-owned, self-operated, and self-managed cruise ship. The company was established in 2014 in Hong Kong hanging Panama flag. The "Splendida" of Diamond Cruises under Taihu International Travel Group is known as the domestic luxury cruise ship. Diamond Cruises is the only private cruise company for

Table 4 Situation of China's domestic cruise brands

Domestic cruise ship	Company	Nationality	Operating state	Total tonnage (10,000 tons)	Standard capacity (people)
Henna	HNA Tourism	Malta	Out of service	4.7	1965
New Century	SkySea Cruise Line	Malta	In service	7.1545	1814
China Taishan	Bohai Ferry	Panama	In service	2.45	900
Diamond Splendida	Diamond Cruises	Bahamas	In service	2.4782	1099
Nanhai Dream	Nanhai Cruises	China	In service	2.4572	893

self-operation, self-management and independent sales with independent ships. The "Nanhai Dream" was built by China's first shipbuilding company, Guangzhou Shipyard International Company Limited. It is transformed from a passenger ship, and now operated and managed by Sansha Nanhai Dream Cruise Limited Company. With Xisha route as the starting point, the South China Sea islands, the Taiwan Strait and other coastal routes are developed vigorously in order to create quality routes meeting the needs of Chinese tourists.

3.2 Analysis on the Operating Pressure Faced by China's Domestic Cruise Brands

In November 2015, "Henna" retired, while "New Century", "China Taishan" and "Diamond Splendida" are still in operation. SkySea Cruise Line is undoubtedly the best "domestic cruise company". However, facing with the competition of international giants, it undergoes huge business pressure. The "New Century" is also called the "national cruise". As the most praiseworthy domestic cruise ship, it is named "the best domestic cruise ship". As it is mainly managed by the Royal Caribbean, the internal decoration is still with an international style; apparently, it is not the "domestic cruise" in a true sense, but a "localized cruise". The route of domestic cruise companies is also along the "Japan and South Korea route". Under the positive competition of foreign international cruise giants, it has no obvious differentiated advantage, making "cost efficiency" becomes the biggest selling point of domestic cruise companies.

With foreign cruise companies occupying the Chinese market, the domestic cruise companies face with many problems, such as small tonnage, second-hand cruise, low brand value, weak financial strength, weak operation and management capacity, and lack of cruise talent. These problems cannot be solved in a short time.

However, China enjoys an absolute advantage in cruise ticket distribution channels, which is hammered by the "charter mode". The price war is extremely hot, making the cruise ticket prices gradually reduced. The "price war" and "charter mode" have also caused a certain impact on foreign cruise companies, but the impact is more intense for domestic cruise companies and travel agents, leading to a sharp decline in their income. Foreign cruise companies can obtain gains from other foreign markets to make up for losses, which increase their competitiveness compared with domestic cruise companies.

3.3 Dilemma Faced by Domestic Cruise Operation

3.3.1 Dilemma of Purchase Tax of Domestic Cruise Ships

In the past few years, in order to promote the expansion of China's cruise market, China has developed a number of favorable policies on the cruise market for equal treatment of all cruise companies and cultivation of the cruise market, but there was not any special support for domestic cruise companies, which has been developed in a relatively slow rate. Up to now, China doesn't have a cruise ship registered in China to hang the five-star red flag. Most ships are registered in Panama or Bahamas and hanging the flag of convenience like many foreign cruise ships, which is correlated with the nearly 30% tax. In foreign countries, the acquisition and registration of ships is subject to an import tariff of 9% and a value-added tax of 17%, with a merger tax rate of 27.53%, which will increase the operating costs of domestic cruise companies and greatly enhance the cost for purchasing the cruise ship.

3.3.2 Dilemma of Retirement Age of the Domestic Cruise Ship

As the pioneer of Chinese cruise brand, the domestic cruise companies are to be given more support and encouragement to enhance their competitive advantage in the market. But they also face a lot of restrictions, for example, there is no specific policy for the purchase of cruise ships. They still adopt the policy for cargo and passenger ship with a retirement period of 30 years. "Henna" is thus retired after the expiration of 30 years. A cruise ship has a big difference from a cargo ship. As "the pearl of the shipbuilding industry", a cruise ship is the fruit of sophisticated technology of the shipbuilding industry characterized by a high degree of precision, thus it usually has a longer service age than the general ship. 10–25 years is the "golden age" for cruise operation. The period of 30 years is obviously short for the cruise ship, which can be extended for about 1.5 times. This policy has little restriction on foreign cruise companies. The early retirement of cruise ships is also an increase in the financial pressure on local cruise companies.

3.3.3 Capital Pressure for Purchase of Domestic Cruise Ships

Cruise ship is a kind of high-value luxury ship, and it is thus very expensive. It costs about CNY1 billion to purchase a second-hand 70,000-ton cruise ship. In China, there is no special cruise financial support, and the special fund has not been established. Without strong financial support, it is difficult for Chinese companies to make the decision to buy a cruise ship.

At present, China's enterprises to buy cruise ships are mostly tourism or shipping enterprises, such as "New Century" purchased by the institution jointly established by Ctrip and Royal Caribbean, "China Taishan" purchased by Bohai Ferry Co., Ltd., "Henna" purchased by the HNA Group, and "Diamond Splendida" purchased by Taihu International Travel Group. The labor-intensive tourism industry makes it difficult to obtain strong financial resources. Therefore, China currently levies a tax nearly 30% for the purchase of cruise ships registered in China and hanging with the five-star red flag, which also restricts the domestic enterprises to purchase foreign cruise ships and encourages them to purchase cruise ships with tax relief.

3.3.4 Slow Development of the Cruise Ship Industry in South China Sea

From the experience in the past few years, we can see that domestic cruise companies always faced huge pressure in a positive confrontation with foreign cruise companies. With good marine and tourism resources, the South China Sea region is an important base for the development of the domestic cruise business. At present, only the supply ship and the "Nanhai Dream" are in operation. The South China Sea cruise market is an important development area for the development of domestic cruise business in the future. With Chinese people as the main source of tourists, its mysterious and sovereign properties have a strong appeal for Chinese people.

According to the current development status of "Nanhai Dream", Chinese people have a special patriotic feeling for this place, and eventhough "Nanhai Dream" costs more than Japan and South Korea cruise lines, they are very willing to bear the costs. The current price for travelling by "Nanhai Dream" is about CNY4000-20,000. The South China Sea cruise market is suitable for the operation of small and medium-sized cruise ships. The ecological vulnerability makes it difficult to bear the large-sized cruise ships. In the South China Sea region, only Xisha and other three islands are now open to the outside world, and there is no port for berthing of cruise ships, which is to be vigorously developed in the future. In the future, we aim to not only opening more waters and islands, but also building ports for berthing of cruise ships.

4 Analysis on Development Strategy and Path of Domestic Cruise Brands

4.1 Development Orientation

As a typical modern service industry, cruise tourism is a typical advanced manufacturing industry featured with cruise manufacturing and guided by "service oriented, high-end breakthrough, agglomeration development, coordination and linkage". By constantly improving the port infrastructure and related supporting service system, cultivating cruise consumer market and environment, extending the cruise economy chain, vigorously promoting the integration of industrial, marine, cultural, commercial and other related industries with traditional tourism industries, we are dedicated to fostering Chinese cruise tourism into a hot spot for outbound travelers and key domestic tourism resort to "experience the slow life, enjoy the waterside life", making it one of the national tourism economic base and the strategic goal of the ocean power.

Based on the linkage of cruise ports along domestic gold coastline and relying on domestic cruise enterprises, an international cruise route is introduced with Vietnam in the west, Singapore and Australia, Hawaii in the south, Hawaii and East Pacific in the east, Japan and South Korea in the north. In the future, we will actively explore the global routes and home port routes with internationally renowned home port in the Mediterranean, the Caribbean, Miami and other countries, and build and disseminate Chinese tourism policy and tourism civilization in the world.

4.2 Development Goals

Facing with the domestic economic development policies and forms, development expectations of the domestic cruise market, and the new development pattern of the global cruise industry, the cruise industry needs to further clarify the future development trend, direction and focus, grasp the ever-changing cruise market environment and the development opportunities of the global cruise industry. The strategic development plan of the cruise industry will start from 2020 when "the 13th five-year plan" period is to be completed, and it will be divided into three five-year periods, i.e., 2020–2025, 2025–2030, and 2030–2035. The four subjects, "port", "cruise line", "travel" and "talents", are oriented in the three development target areas to promote the development of the cruise industry. See Table 5.

12 Analysis on the Development Strategy and Path ... 265

Table 5 Proposed targets for development of China's domestic cruise ships

	Near-term targets	Medium-term targets	Long-term targets
Layout of ports and overseas travel destinations	Plan to improve the domestic (coastal) cruise port and visit port, and to plan the access-port cruise tourism resources	Plan to purchase and lease foreign ports with foreign countries as the home port	Purchase and lease tourist destinations (islands)
Cruise travel	Improve the berthing of foreign cruise ships in the form of visit port and home port, and develop the berthing of China's domestic cruise ships with the Chinese port as the home port	On the basis of the near-term targets, achieve the berthing of Chinese domestic cruise ships with the neighboring countries as the home port and the Chinese port as the assess port	On the basis of the recent and medium-term targets, achieve the berthing of Chinese domestic cruise ships with foreign countries as the home port and the Chinese port as the assess port
Shipbuilding	Carry out preliminary cooperation with the placement of procurement order by foreign shipyard as the main channel	After learning from foreign experience, carry out joint efforts in shipbuilding with foreign countries	China builds ships with independent intellectual property rights independently
Operation and management	China's domestic cruise ships shall be managed mainly by the foreign management team, especially foreign management and service personnel, to learn from foreign experience	The management of China's domestic cruise ships is gradually shifted from foreign management personnel and Chinese service personnel to Chinese management personnel and Chinese service personnel	Basically form the management team comprised by Chinese management personnel and foreign service personnel

4.3 Implementation Focus

4.3.1 Localization of Domestic Cruise Design

"Domestic cruise" has a significant difference from "traditional cruise", which means that domestic cruise companies should take a differentiated competitive strategy. At present, although lots of "Chinese customized" cruise ships appear in the market, such as "Ovation of the Seas", Princess Cruises' "Majestic Princess", Norwegian Cruise Line's "Joy" and Dream Cruises' "Genting Dream", there is still a lack of Chinese flavor. With more Chinese characteristics, domestic cruise ships are integrated with elements favored by Chinese people. Domestic cruise ships should not only be unique in the cruise design, but also showcase unique features in the dining, entertainment, leisure and other aspects. The design and manufacture of domestic cruise ships should be integrated with the elements favored by Chinese people, and also retain the western elements to a certain extent. The target of visitors is to experience the cruise culture. According to the love of Buddhist culture and historical buildings by Chinese people, domestic cruise ships should be integrated with cultural landscapes such as temples, monks, Shikumen and Jiangnan ancient water town, which is also an important direction for the development of domestic cruise enterprises.

4.3.2 Themed Domestic Cruise Products

Based on the development characteristics of entertainment and leisure industries, domestic cruise enterprises should combine a variety of industries with domestic cruise ships, and combine the static landscape effects with regional culture. The themed cruise will be an important development trend, for example, the Disney cruise is designed with the wedding theme to provide the whole process of service for the wedding ceremony, Buddhist theme provides visitors with a meditation travel, and food theme makes visitors experience Asian cuisine. Team customers are an important customer base for domestic cruise tourism. It is thus a good plan to build the international maritime conference center, especially for corporate customers, entrepreneurial customers and executives to carry out marine entrepreneur forum, water regimen, etc.

4.3.3 Internationalization of Domestic Cruise Development

Domestic cruise enterprises should not only create "domestic cruise", but also create "international cruise brand", and bear the burden of creating "national cruise brand". At present, Chinese cruise market is mainly dominated by foreign cruise companies. It is urgent to drive the "domestic cruise" in entering the international cruise market, to achieve the goal of foreign cruise port as the "home port" or "visit port" and enter the forefront line in the global cruise market by creating "China's international cruise giants". In addition to creating cruise companies, domestic cruise enterprises should actively acquire foreign cruise port to achieve the layout of Chinese cruise companies in international cruise market. With the strong strength in international mergers and acquisitions, Chinese enterprises are expected to purchase foreign cruise port soon and gradually occupy a dominant position in the global cruise market.

4.3.4 Differentiation of Competition Strategies in Domestic Cruise Market

For the domestic cruise tourism, a differentiated strategy must be implemented, because the cruise tonnage cannot form a competitive advantage compared with the international cruise company. Medium and small-sized cruise ships are good choices because they can choose high-quality market segments. Small-sized cruise ships with luxury facilities should target the high-end group, forming a good combination with yacht and other high-level tourism ways. With a main focus on the high-income group, they should vigorously develop entrepreneurs, business executives, stars and other customers in order to make full use of island resources. Medium-sized cruise ships are oriented to the middle and high-end group market with a main focus on corporate customers. They should vigorously promote business attributes and carry out customized cruise tourism. Besides, the themed cruise tourism is also a main focus, such as themes of corporate awards, health care topics, historical and cultural experience, Chinese cuisine, high-end fashion, and extreme technology. Large-sized cruise ships are mainly for the mass, especially the youth. The elderly ones are the main force of Japan and South Korea international cruises. The youth is an important factor for the large-sized cruise ships in the domestic cruise market, with a main focus on Chinese fashion, providing offshore "oversea high-speed rail" cruise travel, and a staged trip is offered.

4.3.5 Domestic Cruise Brand Talents in China

The layout of an industry is inseparable from a strong team. Currently, China lacks professional cruise talents. Talent training is a relatively long process. The domestic cruise market not only needs the cruise operation and management personnel, but also needs cruise service personnel. The cruise talent has a strong theoretical and practical knowledge base, and sometimes practice is far more important than theory. So, appropriate adjustment should be made in the theoretical and practical study. The training of cruise talents should not be the cultivation of "all-around talents ", but that of the "professional talents". The targeted training will cultivate a talent with "mature theory, exquisite business, strong service, good practice", so that cruise talents can become professionals upon graduation. In addition, it is necessary to pay attention to the introduction of cruise tourism talents. At present, the international cruise company has attracted a large number of senior professionals by offering generous treatments, such as providing houses, offering education for their children, providing work opportunities for their spouses and career development, making it a "senior cruise talent trainer" in the domestic cruise industry.

5 Conclusion and Outlook

To develop domestic cruise ships, it is necessary to have a Chinese cruise fleet, and design a cruise route suitable for China. The differentiated strategy is to be implemented for the development of domestic cruise ships. Differentiation should not only reflect in the design of hardware of cruise ships with Chinese cultural elements and the cruise product and service with the characteristics of oriental culture, but also in the cruise route design and planning. At present, China has entered into the golden age for the development of international cruise tourism. As it is a "mass tourism era", the tourism consumption will gradually increase worldwide, providing China with more opportunities for developing domestic cruise ships. China is expected to become a "new global cruise center" in the future. China's domestic cruise enterprises will continue to rise. The world cruise market will not only be dominated by the "three giants". In the study on the development trend of the world cruise industry, one can expect the trend of the world cruise industry from the Chinese cruise market. In the planning for the domestic cruise market, the South China Sea region has become the focus of attention, and is considered the golden waterway for the development of China's cruise tourism. Benefiting from the majestic and magnificent natural scenery, the South China Sea region will become an important destination for the development of domestic cruise market, and also an important link for the implementation of differentiated strategy among domestic cruise enterprises.

References

Chapter 3

Li, X. (2016). The relationship between cruise culture and the development of cruise industry at the horizon of Maritime Silk Road. *China Maritime Law, 1*.

Li, X. (2016). Building the 21st century Maritime Silk Road—A tentative discussion on the establishment of a cruise FTA for China, Japan and South Korea. *Economic Review, 10*.

Lu, X., Fu, L., & Zhuang, B. (2017). The study on the potential development of cruise industry cluster in the core area of Hays Road. *Science in the Channel, 5*.

Malong, J., Xiao, H., Chen, S., & Zhu, Z. (2015). Study on Xiamen's cruise economy development from the perspective of one Belt and One Road. *China Water Transport (Second Half), 10*.

Ran, X., Sun, L., & Crab, H. M. (2016). Research on Shanghai cruise line economic contribution based on input-output method. *China Market, 38*.

Roachin. (2014). Fujian to speed up the port development and construction Hays strategic hub. *Fujian Textile, 8*.

Wu, X. (2017). The cruise industry contributes to the development of Maritime Silk Road. *Tourism Overview (Second Half), 4*.

Xu, J. (2014). Depth analysis of the operation of Shanghai cruise home market. *China Port, 6*.

Yao, D. (2016). An exploration into the skills of cruise tour talent based on the perspective of an employable perspective in the Belt and Road. *China Market, 21*.

Zhang, Y. (2017a). Belt and Road under the vision of Sanya cruise tourism development strategy. *Times Finance, 12*.

Zhang, Y. (2017b). Research on the competitiveness of Sanya Phoenix Island cruise port under the Belt and Road initiative. *Tourism Overview (Second Half), 5*.

Zou, L. (2016). Strategies and measures for further Docking Belt and Road in Shanghai. *Shanghai Economic Research, 11*.

Chapter 4

Chase, G. L. (2001). *The economic impact of cruise ships in the 1990s: Some evidence from the Caribbean*. Ph.D., Kent State University, 150.

Dumana, T., & Mattilab, A. S. (2005). The role of effective factors on perceived cruise vacation value. *Tourism Management, 26*, 311–323.

Guo, X. (2013). China cruise tourism research. *Special Zone Economy, 9*.

Henthorpe, T. (2000). An analysis of expenditures by cruise ship passengers in Jamaica. *Journal of Travel Research*, 246–250.

Mackay, C. (2003). *Are we having fun yet? An ethnographic study of a group cruise tour*. Ph.D., The Pennsylvania State University, 242.

Mamoozadedeh, G. A. (1989). *Cruise ships and small island economies: Some evidence from the Caribbean region*. Kent State University, 195.

Mancini, M. (2000). *Cruising. A guide to the cruise line industry* copyright 2000 by Delmar (Vol. 21, pp. 45–47). A division of Thomson Learning.

Qu, H., Ping, E., & Wong, Y. (1999). A service performance model of Hong Kong cruise travelers motivation factors and satisfaction. *Tourism Management, 20*, 237–244.

Wang, H., et al. (2015a). *Report on the development of China's Cruise Industry (2014)*. Social Science Academic Press.

Wang, H., et al. (2015b). *Report on the development of China's Cruise Industry (2015)*. Social Science Academic Press.

Wynen, N. H. (1991). *A survey of the cruise ship industry, 1960-1990* (Vol. l09). Boca Raton, FL: M. A. Florida Atlantic University.

Xie, E, Li, Y. (2003). Review of value creation research based on resource perspective alliance. *Journal of Management Science, 6*.

Xu, H., & Gao, L. (2010). A cruise tour based on supply chain perspective. *Journal of Beijing International Studies University, 1*.

Zhou, H. (2013). Overview of the research on domestic cruise travel. *Oriental Corporate Culture, 20*.

Chapter 5

Brida, J. G., & Zapata, S. (2010). Cruise tourism: Economic, socio-cultural and environmental impacts. *International Journal of Leisure and Tourism Marketing, 1*(3), 205–226.

Chen, L., & Gao, T. M. (1994). Analysis and prediction of macroscopic economic trend by use of the Stock-Watson Prosperity Index. *The Journal of Quantitative & Technical Economics, 5*, 53–59.

Chen, L. Y., Su, R. B., & Li, C. F. (2014). Analysis on the composite index of current Chinese economic prosperity. *Contemporary Economic Research, 2*, 48–53.

Chen, X. Y. (2011). Summary of cruise economy and current domestic research situation. *Economic Vision, 29*, 115–116.

Dai, B., & Group, P. (2008). Study on the cycle index for China's hotel industry. *Journal of Beijing International Studies University, 30*(5), 1–6.

Dai, B., Yan, X., & Huang, X. (2007). Indexed research on the prosperity cycle of Chinese travel agency industry. *Tourism Tribune, 22*(9), 35–40.

Dai, B., Zhang, G. S., Shen, F., & Group, P. (2008). Study on the cycle index for China's hotel industry. *Journal of Beijing International Studies University, 30*(5), 1–6.

Dong, G. Z. (2006). *Study on spatial system of cruise economy*. Zhongshan University.

Dong, W. Q., Guo, T. X., & Gao, T. M. (1987). Determination, analysis and prediction of domestic economic cycle (I)—Existence and determination of economic cycle. *Jilin University Journal Social Sciences Edition, 3*, 1–8.

Gao, M. X. (2014). Theoretical research and empirical analysis on urban tourism prosperity index. Beijing Jiaotong University.

Hai-Min, W. U. (2007). The application of BP-artificial neural network theory in the forecast of Chinese industrial business climate index. *Journal of Lanzhou Commercial College, 23*(2), 84–90.

He, Y., & Zhang, Y. J. (2014). Research on the establishment of tourism prosperity index of Hainan Province. *Economic Research Guide, 1*, 257–258.

Hung, K., & Petrick, J. F. (2012). Testing the effects of congruity, travel constraints, and self-efficacy on travel intentions: An alternative decision-making model. *Tourism Management, 33*(4), 855–867.

Lei, P. (2009). On the drawing up and application of the business index of China's inbound tourism market. *Tourism Tribune, 24*(11), 36–41.

Ni, X., & Dai, B. (2007). Evaluation and analysis research of composite index on tourist industry of China. *Journal of Beijing International Studies University, 11*, 1–4.

Papathanassis, A., & Beckmann, I. (2011). Assessing the poverty of cruise theory hypothesis. *Annals of Tourism Research, 38*(1), 153–174.

Peisley, T. (1999). The cruise business in Asia Pacific. *Travel and Tourism Analyst, 2*, 1–20.

Shen, S. (2011). Overview and prospect of research on China's cruise industry. *Tourism Research, 03*(3), 22–29.

Smith, A. J., Scherrer, P., & Dowling, R. (2009). Impacts on aboriginal spirituality and culture from tourism in the coastal waterways of the Kimberley region, North West Australia. *Journal of Ecotourism, 2*, 82–98.

Sun, X. D. (2015). Chinese cruise tourism industry: New normal and trend. *Tourism Tribune, 30*(1), 10–12.

Wang, X. F. (2010). An empirical research of China's tourism prosperity index. *Statistical Education, 11*, 55–60.

Yang, X. J., & Liu, J. M. (2001). Its types and spatial structure: In the urban tourist development. *Journal of Northwest University, 31*(2), 179–184.

Yu, Y., Chen, J. B., & Feng, D. D. (2015). A composite indicators model of coal industry cycle fluctuation in China. *Resources Science, 37*(5), 969–976.

Zhang, H. (2005). *Research on the environment, system and mode of the Chinese tourism industry in the transformation period*. Tourism Education Press.

Zhang, X. J. (2008). *Study on economic effect of cruise tourism and its transmission mechanism*. Xiamen University.

Chapter 6

Bao, Q. (2014). Analysis of passenger satisfaction of cruise industry—Take the star pisces of star cruise as an example. *Tourism Overview, 12*, 182.

Cai, X., Niu, Y., & Wei, Z. (2010). Research on the development potential of cruise industry in China. *Development Research, 3*, 62–66.

Cai, E., & Shi, J. (2014). A review of domestic cruise port related research. *World Shipping, 37*(8), 17–20.

Du, X., Li, H., & Wen, Y. (2016). Experience and implications of the onshore power supply facility at the cruise terminal Altona in Hamburg, Germany. *Environment and Sustainable Development, 41*(4), 40–43.

Duan, X. (2013). Research on industrialization of Zhejiang cruise tourism under the background of marine economy. *Territory & Natural Resources Study, 6*, 72–75.

Ge, Y. (2010). On development of international cruise counselor—Strategies on cruise counselor development in China. *Journal of Tianjin University of Commerce, 4*, 56–59, 69.

Guan, S., & Wu, X. (2014). Survey of Chinese cruise passenger satisfaction. *Industrial & Science Tribune, 13*(14), 127–128.

He, L., Zhu, J., Cai, J., et al. (2016). Research on the cruise pricing strategy based on the Markov forecasting method. *Journal of Shangqiu Normal University, 32*(9), 1–5.

Hu, J., & Chen, J. (2004). An economic analysis of the development of Shanghai's cruise industry. *Tourism Tribune, 19*(1), 42–46.

Huang, D., & Qiu, L. (2014). Research and analysis on the characteristics of Chinese tourists' cruise consumption behavior. *Shanghai Enterprise, 6*, 78–80.

Kong, J., & Liu, L. (2015). Research on improving international cruise crew service on basis of cruise passenger satisfaction. *Journal of Wuhan Polytechnic, 3*, 98–100.

Li, H., & Zhou, X. (2015). Study on cruise economy development of estuarine harbor city—Taking Nanjing as an example. *World Regional Studies, 24*(1), 113–122.

Lin, S. (2015). Study on consumption behavior and motivation of domestic cruise tourists. *Tourism Overview, 11,* 218–220.

Liu, Y., & Meng, S. (2017). A survey on the consumer behavior of Xiamen cruise ship tourism market. *Journal of Fuqing Branch of Fujian Normal University, 2,* 89–96.

Liu, R., Yang, P., Zhang, L., et al. (2016). Pricing strategies of the cruise based on multiple prediction indices. *Journal of Hebei North University (Natural Science Edition), 32*(7), 56–60.

Luan, H. (2008). *Study on the contribution of the cruise ship port to regional economy.* Dalian Maritime University.

Luan, C. (2016). Cruise liner transportation emergency and its cohesion with legal system. *China Journal of Maritime Law, 27*(1), 47–54.

Ma, D. (2008). A brief discussion on the cultivation of the economic legal issues of the cruise ship in Bohai sea area. *Journal of Dongbei University of Finance and Economics, 6,* 68–71.

Nie, L., & Dong, G. (2010). A study on competitiveness of cruise tourism in port cities based on entropy-TOPSIS. *Tourism Forum, 6,* 789–794.

Qiao, Y. (2010). Analysis of Shanghai cruise tourism marketing strategy. *Modern Business Trade Industry, 1,* 113–114.

Qiu, L., & Xia, X. (2017). Comparative study on the consumption behavior of cruise tourists in China and foreign countries. *Public Utilities, 4*(4), 17–22.

Shao, L., & Zhang, L. (2007). On ways of raising inspecting efficiency for cruise ships. *Journal of Shanghai Police College, 2,* 76–79, 94.

Shen, R., & Min, D. (2007). Learning on cruise-ship culture. *Research on Waterborne Transportation, 1,* 24–26.

Shen, S. (2011). Overview and prospect of the research on China's cruise industry. *Tourism Research, 03*(3), 22–29.

Sun, L., & Wang, C. (2009). Cruise manufacturing—A new star in the future of civil shipbuilding. *World Shipping, 32*(1), 64–65.

Sun, L., & Zhou, Q. (2017). The innovative development of Guangxi cruise tourism in the background of "One Belt And One Road"—Based on the investigation and research on the consumption demand of guangxi cruises. *Social Scientist, 7,* 118–122.

Sun, R., Ye, X., & Xu, H. (2016). Low Price Dilemma in China cruise market: Analysis on the price formation mechanism. *Tourism Tribune, 31*(11), 107–116.

Sun, X., & Feng, X. (2012). Cruise tourism industry in China: Present situation of studies and prospect. *Tourism Tribune, 27*(2), 101–112.

Sun, X., & Feng, X. (2013). How to set prices for cruise cabins: An empirical study on the North America Market. *Tourism Tribune, 28*(2), 111–118.

Sun, X., & Hou, Y. (2017). Evaluation of tourist satisfaction with cruise Homeport: An empirical study on Shanghai. *Scientia Geographica Sinica, 37*(5), 756–765.

Sun, X., & Ni, R. (2017). Onboard attributes/criteria of cruise ships and cruises' satisfaction evaluation. *Statistics & Information Forum, 32*(10), 116–122.

Sun, X., Wu, X., & Feng, X. (2015). Basic characteristics and key elements of cruise itinerary planning. *Tourism Tribune, 30*(11), 111–121.

Sun, Y. (2017). Research on the driving effect of international cruise homeport to regional economy: A case study of Sanya. *Modern Urban Research, 4,* 120–124.

Wang, W., & Zhang, W. (2008). Planning and design of cruise ship homeport. *Port & Waterway Engineering, 12,* 88–93.

Wang, J., Zhou, Q., & Yang, K. (2012). Research of cruise terminal emergency evacuation strategy base on cellular automaton. *Journal od Wuhan University of Technology (Transportation Science & Engineering), 36*(3), 587–589, 593.

Xie, F., Li, H., & Li, D. (2010). Cruise environment pollution control mechanism and tactics based on life-cycle assessment. *Marine Science Bulletin, 29*(6), 702–706.

Xie, L., Zhao, B., & Chen, Y. (2012). Cruise marinas layout plan in Guangdong province. *Port & Waterway Engineering, 5,* 65–67.

Xu, H., & Gao, L. (2010). The cruise tourism based on the perspective of supply chain. *Journal Beijing International Studies University, 1,* 58–62.

Xu, T., Sun, L., & Li, T. (2008). Study on sediment siltation for international mail steamer terminal project of Tianjin Port. *Journal of Waterway and Harbor, 29*(2), 100–105.

Xu, C., & Wu, X. (2015). A survey on Shanghai cruise tourists behavior. *Co-operative Economy & Science, 3,* 43–44.

Yang, J. (2015). Review and prospect of cruise tourism research commentary on the English journal articles. *World Regional Studies, 24*(1), 130–139.

Yang, Y., & Wu, X. (2011). A study on the supply characteristics of cruise tourism market in China. *Special Zone Economy, 9,* 164–167.

Ye, X., & Sun, R. (2007). Research on cruise industry of tourism in Shanghai based on customer demand. *East China Economic Management, 21*(3), 110–115.

Yu, D. (2008). *Research on the competitiveness of Dalian cruise home port.* Dalian Maritime University.

Yu, W., & Huang, J. (2015). The pollution prevention and control of cruise tourism in China and abroad and the enlightenment of hainan in circular economy mode. *Journal of Green Science and Technology, 10,* 296–299.

Zhao, L. (2009). Research on the training mode of cruise tourism talents in colleges and universities. *Maritime Education Research, 26*(2), 57–60.

Zhang, X. (2008). *Research on the economic effect of cruise tourism and its conduction mechanism.* Xiamen University.

Zhang, S., & Cheng, J. (2012). Study on the countermeasures for the development of China's cruise tourism industry. *Tourism Tribune, 27*(6), 79–83.

Zhang, F., & Fang, B. (2006). Prospect analysis of cruise tourism in Qingdao. *China Water Transport (Theory Edition), 4*(6), 26–28.

Zhang, Y., Kou, M., & Ma, B. (2012). A study review on overseas cruise tourism market. *Tourism Tribune, 27*(2), 94–100.

Zhang, W., & Luo, Z. (2011). Progress about the study of overseas cruise tourism. *Journal of Hunan University of Commerce, 18*(5), 77–83.

Zhou, H. (2013). Overview of domestic cruise tourism research. *Oriental Enterprise Culture, 20,* 169–170.

Zhu, L. (2010). Evaluation of cruise ports' tourism competitiveness based on factor analysis. *Journal of Huaihai Institute of Technology (Humanities Forum, Social Science Edition), 8*(9), 40–42.

Chapter 7

Braun, B. M., Xander, J. A., & White, K. R. (2002). The impact of the cruise industry on a region's economy: A case study of Port Canaveral, Florida. *Tourism Economics, 8*(3), 281–288.

Brida, J. G., & Zapata, S. (2010). Economic impacts of cruise tourism: The case of Costa Rica. *Anatolia: An International Journal of Tourism and Hospitality Research, 21*(2), 322–338.

Chase, G., & Alon, I. (2002). Evaluating the economic impact of cruise tourism: A case study of Barbados. *Anatolia: An International Journal of Tourism and Hospitality Research, 13*(1), 5–18.

Chase, G. L., & McKee, D. L. (2003). The economic impact of cruise tourism on Jamaica. *The Journal of Tourism Studies, 14*(2), 16–22.

China National Tourism Administration Department of Statistical Policies and Legislation. (2014). *Tourism ssampling survey 2014.* Beijing: China Travel and Tourism Press.

CLIA. (2014a). *The contribution of the North American Cruise Industry to the U.S. Economy in 2013.* Exton, PA: Business Research & Economic Advisors.

CLIA. (2014b). *The global economic contribution of cruise tourism 2013*. Exton, PA: Business Research & Economic Advisors.
CLIA—North West & Canada. (2013). *The economic contribution of the international cruise industry in Canada 2012*. Exton, PA: Business Research & Economic Advisors.
CLIA Europe. (2014). *Contribution of cruise tourism to the economies of Europe 2014*. Brussels: CLIA Europe.
Dwyer, L., & Peter, F. (Eds.). (2006). *International handbook on the economics of tourism*. Cheltenham: Edward Elgar.
Greater Victoria Harbour Authority. (2013). *The economic contribution of cruise tourism in victoria 2012*. Exton, PA: Business Research & Economic Advisors.
Hawaii DBEDT. (2004). *2002 and 2003 Hawaii cruise industry impact study*. Honolulu: Hawaii DBEDT.
Larry, D., & Peter, F. (1996). Economic impacts of cruise tourism in Australia. *The Journal of Tourism Studies, 7*(2), 36–43.
Larry, D., & Peter, F. (1998). Economic significance of cruise tourism. *Annals of Tourism Research, 25*(2), 393–415.
London Development Agency. (2009). *An assessment of current and future cruise ship requirements in London*. London: LDA.
Mescon, T. S., & Vozikis, G. S. (1985). The economic impact of tourism at the port of Miami. *Annals of Tourism Research, 12*(4), 515–528.
National Bureau of Statistics of China Department of National Accounts. (2011). *China regional input-output tables 2007*. Beijing: China Statistics Press.
NYCruise. (2014). *NYCruise 2013 economic impact study*. New York: NYCruise.
Papathanassis, A., Lukovic, T., & Vogel, M. (Eds.). (2012). *Cruise tourism and society: A socio-economic perspective*. Berlin: Springer.
Port of Los Angeles. (2009). *Cruise market demand evaluation study 2009*. Los Angeles: Port of Los Angeles.
Shanghai Municipal Statistics Bureau, and Survey Office of the National Bureau of Statistics in Shanghai. (2014). *Shanghai statistical yearbook 2014*. Beijing: China Statistics Press.
Smeral, E. (2006). Tourism satellite accounts: A critical assessment. *Journal of Travel Research, 45*(1), 92–98.
Stabler, M. J, Papatheodorou, A., & Thea Sinclair, M. (2010). *The economics of tourism* (2nd ed.). Oxford: Routledge.
UNWTO. (2010). *Cruise tourism—Current situation and trends*. Madrid: UNWTO.
Van Balen, M., Dooms, M., & Haezendonck, E. (2012). *The economic impact of river tourism on ports: The case of Brussels*. Paper presented at International Association of Maritime Economists (IAME) Conference (Vol. 9, pp. 5–8), Taipei.
Vanhove, N. (2011). *The economics of tourism destinations* (2nd ed.). London: Elsevier.
Worley, T., & Akehurst, G. (2013). *Economic impact of the New Zealand cruise sector*. Auckland: Cruise New Zealand.
Wright, A. (2001). *The economic impact of cruise ports: The case of Miami*. Geneva: UNCTAD.
WTTC. (2014). *WTTC/Oxford economics travel & tourism economic impact research* (p. 2014). London: WTTC.

Chapter 8

2013–2014 China cruise market development report. Shanghai International Shipping Research Center, Cruise Economy Research Institute, 2014.
Chen, X. (2016). *Cruise port emergency risk assessment and early warning system*. Shanghai University of Engineering Science.
Cheng, J. (2010). Study on strategy of cruise industry development in China. *World Shipping, 33*(12), 22–24.

References

Dong, G. (2012). Research on Shanghai's countermeasures to improve international sailing ship bonded oil market. *Scientific Development, 10,* 90–104.
Gan, S. (2013). Study on Hanghai cruise industry chain development. *Transportation Enterprise Management, 28*(6), 14–16.
Gao, Q. (2011). *The major legal issues involved in the international sailing ship supply industry in our country.* Fudan University.
Guo, P. (2016). A study on the legal system protection for promotion the development of cruise industry. *Law Science Magazine, 37*(8), 48–54.
Jin, B. (2016). *Made in China 2025.* CITIC Publishing House.
Lau, Y. Y., & Tam, K. (2014). Cruise terminals site selection process: An institutional analysis of the Kai Tak Cruise Terminal in Hong Kong. *Research in Transportation Business & Management, 13,* 16–23.
Liu, H. & Li, W. (2007). The conception of building the home port of cruise ship in Tianjin harbor, *Port Economy, 6,* 32–33.
Ma, Y. & Shi, J. (2009). Construction scale of post ship bus port in Shekou passenger port area, Shenzhen. *Shipping Management, 4,* 14–16.
Mai, Y. et al. (2009). Construction of Wu Song Kou cruise terminal to promote the development of Shanghai. *China Ports, 11,* 31–33.
Miao, H., & Zhu, Y. (2013). A study on influencing factors of the development of China's cruise industry. *China Ports, 2,* 49–53.
Qi, Y., & Shan, D. (2010). A study on evaluation of port cites' cruise tourism industry of competitiveness-also a study on the development of sanya cruise tourism industry. *Research on the Generalized Virtual Economy, 6,* 35–41.
Qiu, L., & Gao, C. (2015). Based on the value chain of the cruise industry value-added effect of research. *Social Sciences in Hunan, 2,* 134–137.
Sun, X. (2015). China's cruise industry: New normals and new trends. *Tourism Tribune, 30*(1), 10–12.
Sun, X., Feng, X., & Gauri, D. K. (2014). The cruise industry in China: Efforts, progress and challenges. *International Journal of Hospitality Management, 42,* 71–84.
Véronneau, S., & Roy, J. (2009). Global service supply chains: An empirical study of current practices and challenges of a cruise line corporation. *Tourism Management, 30*(1), 128–139.
Wang, H., et al. (2016). *Report on development of China cruise industry.* China: Social Sciences Academic Press.
Wu, Y., et al. (2009). A study on the current situation and problems of China's road transport market. *Transport Research, 6,* 71–74.
Wu, Y., & Zhu, Y. (2014). The development of China's cruise industry. *Transportation Enterprise Management, 7,* 12–14.
Xie, X. (2011). Study on China's shipping port supply pattern and development direction. *China Forts, 9,* 11–13.
Xu, J. (2011). Shanghai port shipbuilding industry policy appeal. *China Forts, 9,* 14.
Zeng, Q., & Yuan, S. (2012). A study on the index system of the construction of international cruise terminal. *Jilin Normal University Journal (Natural Science Edition), 11,* 65–68.
Zhang, M., Du, Y., & Sun, J. (2016). Study on the development of Chinese cruise manufacturing. *Ship Engineering,* (s2), 252–254.
Zhang, X., & Zhang, C. (2016). Game analysis on the core competitive advantages of logistics supply chain in cruise industry. *Statistics & Decision, 4,* 58–61.
Zhu, J. (2013). Analysis and suggestions on the development of Shanghai cruise bus port industry. *Shipping Management, 6,* 38–42.
Zhu, B., & Yan, C. (2015). Research on the development and problems of Shanghai cruise ship supply industry. *Communication & Shipping, 2,* 57–61.

Chapter 9

Ahmed, Z. U., Johnson, J. P., et al. (2002). Country-of-origin and brand effects on consumers evaluations of cruise lines. *International Marketing Review, 19*(2/3).

Cai, X., & Niu, Y. (2010). The measurement of the competitive potential of Chinese cruise tourism. *Progress in Geographical Science, 10.*

Cruise Lines International Association (CLIA). *2009 cruise industry overview.* http://www.cruising.org, 2010-07-17.

Hong, W., et al. (2014a). *Analysis and trend of the development situation of China's cruise industry in the past 2013-2014 years.* Report on the development of Chinese cruise industry (2014). Social Science Literature Press.

Hong, W., et al. (2014b). Study on the management system and mechanism of Shanghai aims to promote the development of the cruise industry. Report on the development of Chinese cruise industry (2014). Social Science Literature Press.

Huang, H., & Zhu, Y. (2014). *The experience of the world famous cruise city's development and the enlightenment to Shanghai.* Report on the development of Chinese cruise industry (2014). Social Science Literature Press.

Ji, W. (2012). Study on the mode of foreign consumer credit and its reference to China. *Time Finance, 36.*

Jiang, Q. (2013). The role of the German KG fund in ship financing and its reference to China. *German Research, 1.*

Jiang, R. (2014). Market analysis and development trend of Chinese Cruise Port. *Social Science Literature Press, 10.*

Luan, H. (2008). *Study on the contribution of the cruise ship port to regional economy.* Dalian Maritime University.

Ma, C. (2013). The international division of labor in the cruise industry and the competitive strategy of China. *China Economic and Trade Herald, 26.*

Ma, C. (2014). The development of cruise industry in Europe and America: Characteristics, problems and enlightenment. *China Economic and Trade Herald, 23.*

Ma, L., & Wang, J. (2016). The construction of China's cruise market's financial support system based on industrial chain. *Journal of Yangzhou University (Humanity and Social Science Edition), 1.*

Mao, M., & Liu, Y. (2013). Research on SWOT analysis and financial support strategy of China's shipping industry. *Financial Theory and Practice, 2.*

Nevitt, P. K., & Fabozzi, F. J. (2007). CFA, Equipment leasing. *World Shipping, 27.*

Papathanassis, A., & Beckmann, I. (2011). Assessing the poverty of cruise theory' hypothesis. *Annals of Tourism Research, 38*(1).

Qian, H. (2010). On the current situation of China's shipping finance and the reference of international experience. *Modern commercial trade industry, 12.*

Rui, J. (2014). *Market analysis and development trend of Chinese cruise port.* Report on the development of Chinese cruise industry. Social Science Literature Press.

Soriani, S., Bertazzon, S., et al. (2009). Cruising in the mediterranean: Structural aspects and evolutionary trends. *Maritime Policy and Management, 36*(3).

Sun, X. (2014). Cruise industry and cruise economy. *Shanghai Jiao Tong University Press.*

Sun, X., & Feng, X. (2012). Chinese cruise tourism industry: Present situation and prospect. *Tourism Tribune, 2.*

Sun, X., Jiao, Y., & Tian, P. (2011). Marketing research and revenue optimization for the cruise industry. A concise review. *International Journal of Hospitality Management, 30*(3).

Tao, Y., & Liu, C. (2009). Research on the operating mechanism of China's shipping industry investment fund. *Journal of Jiangsu University of Science and Technology (Social Science Edition), 9.*

Wei, L. (2011). The related experience of foreign ship financing lease. *China Maritime Safety, 4.*

Xiao, W. (2014). On the risk of project financing and its evasion measures. *China Business and Trade, 3*.
Yan, H., Zheng, S., Huang, S., Wang, A., & Zhao, X. (2014). The reference significance of the development of the international offshore financial market to the construction of the Shanghai free trade area. *Shanghai Economic Review, 10*.
Yang, J., & Han, L. (2010). Research on the risks and countermeasures of China's ship financing lease under the global economic crisis. *Science of Science and Management of S&T, 6*.
Yin, X. (2016). The operating characteristics of the three largest cruise ship company in the world. *China Ship Survey, 6*.
Zhang, X. (2016). Hundreds of billions of cruise industry outbreaks: How do Chinese companies share a cup of soup? Financial report of the 21st century, 5.
Zhang, Y., Ma, B., & Fan, Y. (2010). The economic characteristics, development trend of cruise industry and its enlightenment to China. *Journal of Beijing International Studies University, 7*.
Zhang, L., & Zhang, Y. (2016). Application of PPP model in the development and operation of cruise ship port: Review and prospect. *World Shipping, 8*.
Zhao, X. (2011). Ship financing lease: urgent policy support. *China Maritime, 4*.
Zhao, T. (2016). The development strategy of Chinese cruise industry based on industrial chain perspective. *Journal of Suzhou University, 6*.
Zuo, Y. (2015). Capital. Analysis of the prospect of Chinese cruise market. *China Tourism News, 12*.

Chapter 10

City under construction or want cruise terminal. Pearl River Water Transport, 2013(14).
Chen, J. (2016). The dream of manufacturing domestic luxury cruise ships gradually realized. *Guangdong Shipbuilding, 4*.
He, Y., & Wu, X. (2006). *International economic law* (Vol. 1). Northeastern University Press.
Li, X., & Yan, C. (2013). Several policy and legal issues on China's development of the cruise industry. *China Maritime Law Research, 3*.
Lv, F. (2015). *Cruise law from the perspective of transport studies*. Doctoral dissertation, Dalian Maritime University.
Lv, F., & Guo, P. (2014). An economic analysis of the cruise economy from a legal perspective. *Social Science and Letters, 1*.
Peng, Y. (2014). To create a legal environment for the marine economy—Cruise tourism as an example. *New Economy, 31*.
Su, F. (2014). Analysis of the development of China's cruise tourism risk and coping strategies. *Journal of Wuhan Communications Vocational College, 1*.
Zhang, M., & Liu, Y. (2014). Cruise tourism development review. *Baoshan University Journal, 1*.
Zhao, T. (2016). Based on the industrial chain perspective of China's cruise industry development strategy. *Journal of Suzhou University, 6*.

Chapter 11

Adams, A. W. (2010). Planning for cruise ship resilience: An approach to managing cruise ship impacts in Haines, Alaska. *Coastal Management, 38*(6), 654–664.
Ballini, F., & Bozzo, R. (2015). Air pollution from ships in ports: The socio-economic benefit of cold-ironing technology. *Research in Transportation Business & Management, 17*, 92–98.
Butt, N. (2007). The impact of cruise ship generated waste on home ports and ports of call: A study of Southampton. *Marine Policy, 31*(5), 591–598.
Carić, H. (2011). Cruising tourism environmental impacts: Case study of Dubrovnik, Croatia. *Journal of Coastal Research, 61*(61), 104–113.

Carić, H. (2016). Challenges and prospects of valuation—cruise ship pollution case. *Journal of Cleaner Production, 111* Part B, 487–498.

Carić, H., & Mackelworth, P. (2014). Cruise tourism environmental impacts—The perspective from the Adriatic Sea. *Ocean & coastal management, 102* Part A, 350–363.

Cheer, J. M. (2016). Cruise tourism in a remote small Island—High yield and low impact? In *Handbook of cruise ship tourism*. Oxfordshire: CABI.

Copeland, C. (2008). Cruise ship pollution: Background, laws and regulations, and key issues. Congressional research service reports.

Copeland, C. (2011). Cruise ship pollution: Background, laws and regulations, and key issues. Congressional research service reports.

Cruise Lines International Association (CLIA). (2016). Cruise industry outlook. http://www.cruising.org.

Dawson, J., Johnston, M. E., & Stewart, E. J. (2014). Governance of Arctic expedition cruise ships in a time of rapid environmental and economic change. *Ocean & Coastal Management, 89,* 88–99.

Diedrich, A. (2010). Cruise ship tourism in Belize: The implications of developing cruise ship tourism in an ecotourism destination. *Ocean & Coastal Management, 53*(5), 234–244.

Du, X., Li, H., & Wen, Y. (2016). Experience and enlightenment of green shore power in Altona cruise terminal in Hamburg, Germany. *Environmental and Sustainable Development, 41*(4), 40–43.

Eckhardt, S., Hermansen, O., Grythe, H., et al. (2013). The influence of cruise ship emissions on air pollution in Svalbard—a harbinger of a more polluted Arctic?. *Atmospheric Chemistry and Physics, 13*(1), 3031–3093.

Erize, F. J. (1987). The impact of tourism on the Antarctic environment. *Environment International, 13*(1), 133–136.

Fan, Y., & Yin, Y. (2016). Collaborative governance model of cross-border environmental issues—A theoretical discussion and three case. *Chinese Journal of Public Administration, 2,* 63–75.

Gareau, B. J. (2015). Theorizing environmental governance of the world system: Global political economy theory and some applications to stratospheric ozone politics. *Journal of World-Systems Research, 18*(2), 187.

Genç, R. (2016). Sustainability in cruise ship management. *International Journal of Social Science Studies, 4*(6), 78–83.

He, J. (2014). Modern sea power in the framework of international law and China's maritime rights. *Law Review, 1,* 92–99.

Howitt, O. J. A., Revol, V. G. N., Smith, I. J., et al. (2010). Carbon emissions from international cruise ship passengers' travel to and from New Zealand. *Energy Policy, 38*(5), 2552–2560.

Johnson, D. (2002). Environmentally sustainable cruise tourism: A reality check. *Marine Policy, 26*(4), 261–270.

Li, H., & Lv, S. (2016). Cruise tourism environmental pollution measurement and control scenario analysis—A case study of Shanghai Port. *Marine Development and Management, 12,* 32–38.

Maragkogianni, A., & Papaefthimiou, S. (2015). Evaluating the social cost of cruise ships air emissions in major ports of Greece. *Transportation Research Part D: Transport and Environment, 2015*(36), 10–17.

Miller, F. P., Vandome, A. F., & Mcbrewster, J. (2013). *Cruise ship pollution*. Alphascript Publishing.

Mölders, N., Gende, S., & Pirhalla, M. (2013). Assessment of cruise–ship activity influences on emissions, air quality, and visibility in Glacier Bay National Park. *Atmospheric Pollution Research, 4*(4), 435–445.

Morehouse, C., & Koch, D. (2003). Alaska's cruise ship initiative and the Commercial Passenger Vessel Environmental Compliance Program. *Oceans, 1,* 372–375.

Podolsky, G. (2004). Cruise ship blues: the underside of the cruise industry. *15*(2), 151–152.

Poplawski, K., Setton, E., McEwen, B., Hrebenyk, D., Graham, M., & Keller, P. (2011). Impact of cruise ship emissions in Victoria, BC, Canada. *Atmospheric Environment, 45*(4), 824–833.

References

Qiu, C. (2007). Reflections on anti-pollution caused by the cruise economy. *Battle Transport Research, 03,* 54–56.

Ren, C., Sun, R., & Ye, X. (2017). Cruise ship port environmental protection research prospects. *Journal of Shanghai University of Engineering Science, 31*(2), 166–173.

Ritter, W., & Schafer, C. (1998). Cruise-tourism: A chance of sustainability. *Tourism Recreation Research, 23*(1), 65–71.

Rogers, C. S., & Garrison, V. H. (2001). Ten years after the crime: Lasting effects of damage from a cruise ship anchor on a coral reef in St. John, U.S. Virgin Islands. *Bulletin of Marine Science—Miami, 69*(2), 793–803.

United States Environmental Protection Agency. (2008). Cruise ship discharge assessment report, EPA842-R-07-005.

Wang, J., & Huang, J. (2015). Cruise tourism pollution prevention at home and abroad and its revelation in Hainan under the mode of circular economy. *Green Technology, 10,* 296–299.

Wu, G.-Y., & Shen, Y.-B. (2012). Design concept of shoreline design for lare cruise ship terminal. *Water Transport Engineering, 5,* 74–76.

Ye, X. L., Sun, R., & Meng, M. M. (2017). Green cruise port development and break through the path. *China Ship Inspection,* (7), 60–63.

Yin, J. (2016). On the plight and breakthrough of the global environmental governance model. *Foreign Social Sciences, 5,* 75–82.

Zhao, J., & Bai, J. (2012). Study on environmental issues in international cruise terminal operation. *Tianjin Science and Technology, 3,* 95–96.